Lecture Notes in Computer Science 1801

Edited by G. Goos, J. Hartmanis and J. van Leeuwen

Springer
Berlin
Heidelberg
New York
Barcelona
Hong Kong
London
Milan
Paris
Singapore
Tokyo

Julian Miller   Adrian Thompson
Peter Thomson Terence C. Fogarty (Eds.)

Evolvable Systems: From Biology to Hardware

Third International Conference, ICES 2000
Edinburgh, Scotland, UK, April 17-19, 2000
Proceedings

 Springer

Series Editors

Gerhard Goos, Karlsruhe University, Germany
Juris Hartmanis, Cornell University, NY, USA
Jan van Leeuwen, Utrecht University, The Netherlands

Volume Editors

Julian Miller
The University of Birmingham, School of Computer Science
Edgbaston, Birmingham B15 2TT, UK
E-mail: J.Miller@cs.bham.ac.uk

Adrian Thompson
Sussex University, COGS
Brighton BN1 9RH, UK
E-mail: adrianth@cogs.sussex.ac.uk

Peter Thomson
Napier University, School of Computing
Edinburgh EH14 1DJ, UK
E-mail: petert@dcs.napier.ac.uk

Terence C. Fogarty
South Bank University, School of Computing
Information Systems and Mathematics
London, SE1 0AA, UK
E-mail: fogarttc@sbu.ac.uk

Cataloging-in-Publication Data applied for

Die Deutsche Bibliothek - CIP-Einheitsaufnahme

Evolvable systems : from biology to hardware ; third international
conference ; proceedings / ICES 2000, Edinburgh, Scotland, UK, April
17 - 19, 2000. Julian Miller ... (ed.). - Berlin ; Heidelberg ; New
York ; Barcelona ; Hong Kong ; London ; Milan ; Paris ; Singapore ;
Tokyo : Springer, 2000
  (Lecture notes in computer science ; Vol. 1801)
  ISBN 3-540-67338-5
Cover Photo: Andrew Syred/Science Photo Library

CR Subject Classification (1998): B.6, B.7, F.1, I.6, I.2, J.2, J.3

ISSN 0302-9743
ISBN 3-540-67338-5 Springer-Verlag Berlin Heidelberg New York

Springer-Verlag is a company in the BertelsmannSpringer publishing group.
© Springer-Verlag Berlin Heidelberg 2000
Printed in Germany

Typesetting: Camera-ready by author
Printed on acid-free paper   SPIN: 10720157   06/3142   5 4 3 2 1 0

Preface

A dictionary interpretation of an 'Evolvable System' would be "anything formed of parts placed together or adjusted into a connected whole" that can be "gradually worked out or developed." When we then consider our subtitle 'From Biology to Hardware', we are plunged into a wonderful world of possibilities mixing the leading edge of hardware technology with our best understanding of natural and biological processes.

For the majority of the papers in this volume, the 'System' is some sort of electronics that needs designing, and the 'gradual development' is an algorithm capturing some of the essentials of Darwinian evolution. However, our wider definition of 'Evolvable' is partly explored in papers taking some of the abstract principles of nervous systems, immune systems, and of multicellular development and self-repair. There is a fascinating mix of the here-and-now of engineering applications and problem solving, with radical projects for future systems of a kind not seen before. As the field develops, we expect it to include a wider inspiration from nature at large, and other kinds of 'Hardware': for example with mechanical, fluid, few-electron, or chemical components, perhaps designed at nano-scales.

Nature is not a box of ready-made solutions to fulfil our engineering aspirations. Within these pages, great creativity can be seen in ingeniously transducing and adapting ideas from nature for our technological ends. Many of the papers contribute to a general understanding of how this can work, rather than showing how one specific problem can be solved.

This was the third international conference of the series started in 1996 (with an earlier workshop in 1995; see volumes 1062, 1259, and 1478 of the Lecture Notes in Computer Science series). We thank the scrupulous reviewers, and above all the authors, for enabling this collection of papers to be of a quality that can be viewed as a landmark in this nascent field.

We are very grateful to Xilinx Inc., Cyberlife, and the EvoElec working group of EvoNet for their generous financial support. We much appreciate the IEE's promotion of the conference. Finally we thank whole-heartedly the University of Sussex, the University of Birmingham, and Napier University.

Enjoy!

April 2000 Julian Miller, Adrian Thompson
Peter Thomson, and Terence C. Fogarty

Organization

ICES 2000 was organized by the ICES International Steering Committee.

International Steering Committee

Terence C. Fogarty (South Bank University, UK)
Tetsuya Higuchi (Electrotechnical Laboratory, Japan)
Hiroaki Kitano (Sony Computer Science Laboratory, Japan)
Julian Miller (University of Birmingham, UK)
Moshe Sipper (Swiss Federal Institute of Technology, Switzerland)

Adrian Thompson (University of Sussex, UK)

Organizing Committee

General chair: Terence C. Fogarty (South Bank University, UK)
Program co-chair: Julian Miller (University of Birmingham, UK)
Program co-chair: Adrian Thompson (University of Sussex, UK)
Local chair: Peter Thomson (Napier University, UK)

Program Committee

David Andre, University of California at Berkeley, USA
Juan Manuel Moreno Arostegui, University Polytechnic of Catalonia, Spain
Wolfgang Banzhaf, University of Dortmund, Germany
Gordon Brebner, University of Edinburgh, UK
Stefano Cagnoni, University of Parma, Italy
Prahabas Chongstitvatana, Chulalongkorn University, Thailand
Carlos A. Coello, LANIA, Mexico
Fulvio Corno, Turin Polytechnic, Italy
Peter Dittrich, University of Dortmund, Germany
Marco Dorigo, Free University of Brussels, Belgium
Rolf Drechsler, Albert-Ludwigs-University, Germany
Marc Ebner, Eberhard-Karls-University Tubingen, Germany
Martyn Edwards, UMIST, UK
Stuart J. Flockton, Royal Holloway, UK
Dario Floreano, Swiss Federal Institute of Technology, Switzerland
Terence C. Fogarty, South Bank University, UK
David B. Fogel, Natural Selection, Inc., USA
Max Garzon, University of Memphis, USA

Darko Grundler, Univesity of Zagreb, Croatia
Alister Hamilton, Edinburgh University, UK
Ken Hayworth, Jet Propulsion Lab, USA
Enrique Herrera, University of Granada, Spain
Jean-Claude Heudin, Pôle University Léonard de Vinci, France
Tetsuya Higuchi, Electrotechnical Laboratory, Japan
Tomofumi Hikage, NTT, Japan
Tony Hirst, Open University, UK
Lorenz Huelsbergen, Bell Labs, Lucent Technologies, USA
Hitoshi Iba, The University of Tokyo, Japan
Norbert Imlig, NTT, Japan
Masaya Iwata, ETL, Japan
Didier Keymeulen, Jet Propulsion Laboratory, USA
Tatiana Kalganova, Napier University, UK
Yukinori Kakazu, Hokkaido University, Japan
Hiroaki Kitano, Sony Computer Science Laboratory, Japan
John R. Koza, Stanford University, USA
William B. Langdon, CWI, The Netherlands
Paul Layzell, University of Sussex, UK
Yun Li, University of Glasgow, UK
Henrik Hautop Lund, University of Aarhus, Denmark
Evelyne Lutton, INRIA, France
Patrick Lysaght, University of Strathclyde, UK
Daniel Mange, Swiss Federal Institute of Technology, Switzerland
Pierre Marchal, CSEM, Switzerland
Jean-Arcady Meyer, Ecole Normale Superieure, France
Julian Miller, University of Birmingham, UK
Claudio Moraga, University of Dortmund, Germany
Jan J. Mulawka, Warsaw University of Technology, Poland
Masahiro Murakawa, University of Tokyo, Japan
Kazuyuki Murase, Fukui University, Japan
Norberto Eiji Nawa, ATR, Japan
Andres Perez-Uribe, Swiss Federal Institute of Technology, Switzerland
Marek Perkowski, Porland Center for Advanced Technology, USA
Riccardo Poli, University of Birmingham, UK
Edward Rietman, Bell Laboratories, USA
Hidenori Sakanashi, ETL, Japan
Mehrdad Salami, Swinburne University of Technology, Australia
Eduardo Sanchez, Swiss Federal Institute of Technology, Switzerland
Ken Sharman, University of Glasgow, UK
Moshe Sipper, Swiss Federal Institute of Technology, Switzerland
Giovanni Squillero, Turin Politechnic, Italy
Adrian Stoica, Jet Propulsion Lab, USA
Uwe Tangen, GMD, Germany
Andrea G. B. Tettermanzi, Milan University, Italy

Adrian Thompson, Sussex University, UK
Peter Thomson, Napier University, UK
Marco Tomassini, University of Lausanne, Switzerland
Jim Torrensen, University of Oslo, Norway
Brian Turton, University of Wales, UK
Andy Tyrrell, University of York, UK
Vesselin Vassilev, Napier University, UK
Lewis Wolpert, University College London, UK
Roger Woods, Queen's University of Belfast, UK
Xin Yao, Australian Defense Force Academy, Australia
Ali M S Zalzala, Heriot-Watt University, UK
Ricardo Salem Zebulum, Jet Propulsion Laboratory, USA

Sponsoring Institutions

Xilinx Inc.
Cyberlife
EvoNet: the Network of Excellence in Evolutionary Computing
Institute of Electrical Engineers, UK
University of Sussex, UK
University of Birmingham, UK
Napier University, UK

Table of Contents

Automatic Synthesis, Placement, and Routing of an Amplifier Circuit by Means of Genetic Programming

Forrest H Bennett III
Genetic Programming Inc.
(Currently, FX Palo Alto Laboratory, Palo Alto, California)
forrest@evolute.com

John R. Koza
Stanford University, Stanford, California
koza@stanford.edu

Jessen Yu
Genetic Programming Inc., Los Altos, California
jyu@cs.stanford.edu

William Mydlowec
Genetic Programming Inc., Los Altos, California
myd@cs.stanford.edu

Abstract

The complete design of a circuit typically includes the tasks of creating the circuit's placement and routing as well as creating its topology and component sizing. Design engineers perform these four tasks sequentially. Each of these four tasks is, by itself, either vexatious or computationally intractable. This paper describes an automatic approach in which genetic programming starts with a high-level statement of the requirements for the desired circuit and simultaneously creates the circuit's topology, component sizing, placement, and routing as part of a single integrated design process. The approach is illustrated using the problem of designing a 60 decibel amplifier. The fitness measure considers the gain, bias, and distortion of the candidate circuit as well as the area occupied by the circuit after the automatic placement and routing.

1 Introduction

The *topology* of a circuit involves specification of the gross number of components in the circuit, the identity of each component (e.g., transistor, capacitor), and the connections between each lead of each component. *Sizing* involves the specification of the values (typically numerical) of each component. *Placement* involves the assignment of each of the circuit's components to a particular geographic (physical) location on a printed circuit board or silicon wafer. *Routing* involves the assignment of a particular geographic location to the wires connecting the various components.

Design engineers typically perform the tasks of creating circuit's topology, sizing, placement, and routing as a series of four separate sequential tasks. Each of these tasks is either vexatious or computationally intractable. In particular, the problem of placement and the problem of routing (both analog and digital) are computationally

J. Miller et al. (Eds.): ICES 2000, LNCS 1801, pp. 1–10, 2000.

intractable combinatorial optimization problems that require computing effort that increases exponentially with problem size (Garey and Johnson 1979).

The mandatory requirements for an acceptable scheme for placement and routing are that there must be a wire connecting every lead of all of the circuit's components, that wires must not cross on a particular layer of a silicon chip or on a particular side (or layer) of a printed circuit board, and that minimum clearance distances must be maintained between wires, between components, and between wires and components. Once these mandatory requirements are satisfied, minimization of area typically becomes the next most important consideration.

Genetic programming has recently been shown to be capable of solving the problem of automatically creating the topology and sizing for an analog electrical circuit from a high-level statement of the circuit's desired behavior (Koza, Bennett, Andre, and Keane 1996; Bennett, Koza, Andre, and Keane 1996; Koza, Bennett, Andre, and Keane 1999; Koza, Bennett, Andre, Keane, and Brave 1999). Numerous analog circuits have been designed using genetic programming, including lowpass, highpass, bandpass, bandstop, crossover, multiple bandpass, and asymmetric filters, amplifiers, computational circuits, temperature-sensing circuits, voltage reference circuits, a frequency-measuring circuit, source identification circuits, and analog circuits that perform digital functions. The circuits evolved using genetic programming include eleven previously patented circuits.

However, this previous work did not address the problem of automatically placing and routing of components and wires at particular geographic locations on a printed circuit board or silicon wafer. This paper demonstrates that genetic programming can be used to automatically create the topology, sizing, placement, and routing of analog electrical circuits. Section 2 presents our method. Section 3 describes the preparatory steps required to apply our method to an illustrative problem involving designing a 60 decibel amplifier. Section 4 presents the results.

2   Method

A printed circuit board or silicon wafer has a limited number of layers that are available for wires and a limited number of layers (usually one for a wafer and one or two for a board) that are available for both wires and components. Each wire and component is located at a particular relative geographic (physical) location on the printed circuit board or silicon wafer.

We create electrical circuits using a developmental process in which the component-creating functions, topology-modifying functions, and development-controlling functions of a circuit-constructing program tree are executed. Each of these three types of functions is associated with a modifiable wire or modifiable component in the developing circuit. The starting point of the developmental process is an initial circuit consisting of an embryo and a test fixture. The embryo consists of modifiable wire(s). The embryo is embedded into a test fixture consisting of fixed (hard-wired) components (e.g., the source of the incoming signal) and certain fixed wires that provide connectivity to the circuit's external inputs and outputs. Until the modifiable wires are modified by the developmental process, the circuit produces only trivial output. An electrical circuit is developed by progressively applying the functions in a circuit-constructing program tree (in the population being bred by genetic programming) to the modifiable wires of the original embryo and to the modifiable components and modifiable wires created during the developmental process. The functions in the program tree are progressively applied (in a breadth-first order) to the initial circuit and its successors until a fully developed circuit emerges.

2.1   The Initial Circuit

Figure 1 shows a one-input, one-output initial circuit (consisting of an embryo and a test fixture) located on one layer of a silicon wafer or printed circuit board. The embryo consists of the three modifiable wires, Z0, Z1, and Z2 (in the middle of the figure). All development originates from these modifiable wires. The test fixture contains two ground points G, an input point V (lower left), an output point O (upper right), nonmodifiable wires (hashed), and four nonmodifiable resistors. There is a fixed 1 kilo-Ohm (kΩ) source resistor R4, a fixed 1 kΩ load resistor R18, a fixed 1 giga-Ohm feedback resistor R14, and a fixed 999 Ω balancing resistor R3.

Each element of this initial circuit (and all successor circuits created by the developmental process) resides at particular geographic location on the circuit's two-dimensional substrate. Each element occupies a particular amount of space. For example, the resistors each occupy a 3×3 area; the source point V and the output probe point O each occupy a 1×1 area; the nonmodifiable wires each occupy a $1 \times n$ or $n \times 1$ area; the modifiable wires each occupy a 1×1 area.

The initial circuit in the developmental process complies with the requirements that wires must not cross on a particular layer of a silicon chip or on a particular side of a printed circuit board, that there must be a wire connecting 100% of the leads of all the circuit's components, and that minimum clearance distances between wires, between components, and between wires and components must be respected. Each of the circuit-constructing functions (described below) preserves compliance with these requirements. Thus, every fully laid-out circuit complies with these requirements.

The component-creating functions insert a component into the developing circuit and assign component value(s) to the new component.

2.2   Circuit-Constructing Functions

Figure 2 shows a partial circuit containing four capacitors (C2, C3, C4, and C5) and a modifiable wire Z0. Each capacitor occupies a 3×3 area and is located at a particular geographic location (indicated by an X and Y coordinate). Each piece of wire occupies a $1 \times n$ or an $n \times 1$ area. The modifiable wire Z0 occupies a 1×1 area.

Figure 3 shows the result of applying the one-argument transistor-creating NPN-TRANSISTOR-LAYOUT function to the modifiable wire Z0 of figure 2. The newly created *npn* (q2n3904 BJT) transistor Q6 occupies a 3×3 area and is located at (18, 20). The newly created component is larger than that which it replaces. Thus, its insertion affects the locations of preexisting components C2 and C3 in the developing circuit. Specifically, preexisting capacitor C2 is pushed north by one unit thereby relocating it from (18, 23) to (18, 24). Similarly, preexisting capacitor C3 is pushed south by one unit thereby relocating it from (18, 17) to (18, 16). In actual practice, all adjustments in location are made after the completion of the entire developmental process. Details of implementation of this function (and other functions described herein) are found in Koza and Bennett 1999. The newly created transistor is not subject to subsequent modification (and hence there is no construction-continuing subtree).   Similarly, the PNP-TRANSISTOR-LAYOUT function inserts a *pnp* (q2n3906 BJT) transistor.

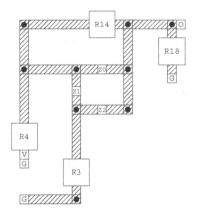

Figure 1 Initial circuit consisting of embryo and test fixture.

The two-argument capacitor-creating LAYOUT-C function inserts a capacitor into a developing circuit in lieu of a modifiable wire (or modifiable component). This component-creating function takes an argument that specifies component sizing. Similar functions insert other two-leaded components (e.g., resistors and inductors). The sizing of components are established by a numerical value. In the initial random generation of a run, the numerical value is set, individually and separately, to a random value in a chosen range. In later generations, the numerical value may be perturbed by a mutation operation using a Gaussian probability distribution.

2.3   Topology-Modifying Functions

The topology-modifying functions modify the topology of the developing circuit.

The two-argument SERIES-LAYOUT function creates a series composition consisting of the modifiable wire or modifiable component with which the function is associated and a copy of the modifiable wire or modifiable component.

Each of the two functions in the PARALLEL-LAYOUT family of four-argument functions creates a parallel composition consisting of two new modifiable wires, the preexisting modifiable wire or modifiable component with which the function is associated, and a copy of the modifiable wire or modifiable component.

The one-argument polarity-reversing FLIP function reverses the polarity of the modifiable component or modifiable wire with which the function is associated.

Most practical circuits are not entirely planar. Vias provide a way to connect distant points of a circuit. Each of the four functions in the VIA-TO-GROUND-LAYOUT family of three-argument functions creates a T-shaped composition consisting of the modifiable wire or modifiable component with which the function is associated, a copy of it, two new modifiable wires, and a via to ground. There is a similar VIA-TO-POSITIVE-LAYOUT family of four three-arguments functions to allow direct connection to a positive power supply and a similar VIA-TO-NEGATIVE-LAYOUT family for the negative power supply.

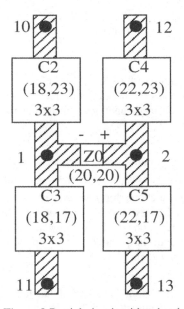

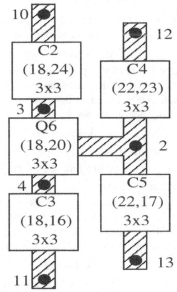

Figure 2 Partial circuit with a 1 × 1 piece of modifiable wire Z0 at location (20, 20).

Figure 3 The result of applying the NPN-TRANSISTOR-LAYOUT function.

Similarly, numbered vias provide connectivity between two different layers of a multi-layered silicon wafer or multi-layered printed circuit board. A distinct four-member family of three-argument functions is used for each layer. For example, the VIA-0-LAYOUT and VIA-1-LAYOUT families of functions makes connection with a layers numbered 0 and 1, respectively, of a two-layered substrate.

2.4   Development-Controlling Functions

The zero-argument END function makes the modifiable wire or modifiable component with which it is associated into a non-modifiable wire or component (thereby ending a particular developmental path).

The one-argument NOOP ("No Operation") function has no effect on the modifiable wire or modifiable component with which it is associated; however, it delays the developmental process on the particular path on which it appears.

3   Preparatory Steps

The method will be illustrated on the problem of creating the topology, component sizing, placement, and routing for a 60 dB amplifier with zero distortion and zero bias and with the smallest possible total area for the bounding rectangle of the fully laid-out circuit. (See Bennett, Koza, Andre, and Keane 1996 for a more detailed statement of this problem, without consideration of placement and routing). The circuit is to be constructed on a two-sided printed circuit board with two internal layers. The top side contains discrete components (e.g., transistors, capacitors, and resistors,) that are connected by perpendicularly intersecting metallic wires. The bottom side is devoted to connections to ground. The two internal layers are devoted to via 0 and via 1.

3.1   Initial Circuit

We use the one-input, one-output initial circuit (figure 1) consisting of a test fixture and an embryo with three modifiable wires.

3.2   Program Architecture

There is one result-producing branch in the program tree for each modifiable wire in the embryo. Thus, the architecture of each circuit-constructing program tree has three result-producing branches. Neither automatically defined functions nor architecture-altering operations are used.

3.3   Function and Terminal Sets

The terminal set, $\mathcal{T}_{ccs}$, for each construction-continuing subtree consists of the development-controlling END function. The function set, $\mathcal{F}_{ccs}$, for each construction-continuing subtree includes component-creating functions for *npn* transistors, *pnp* transistors, capacitors, resistors, and inductors (a totally extraneous component for this problem); the development-controlling NOOP function; and topology-modifying functions for series, parallel, flips and vias to ground, the positive power supply, the negative power supply, and layers 0 and 1 of the printed circuit board.

3.4   Fitness Measure

The fitness measure is based on the area of the bounding rectangle of the laid-out circuit as well as the gain, bias, and distortion of the candidate amplifier circuit.

The evaluation of the fitness of each individual circuit-constructing program tree in the population begins with its execution. This execution progressively applies the functions in the program tree to the embryo of the circuit, thereby creating a fully developed (and fully laid out) circuit. Since the developmental process for creating the fully developed circuit includes the actual geographic placement of components and the actual geographic routing of wires between the components, the area of the bounding rectangle for the fully developed circuit can be easily computed.

A netlist is then created that identifies each component of the developed circuit, the nodes to which each component is connected, and the value of each component. The netlist is the input to our modified version of the SPICE simulator (Quarles, Newton, Pederson, and Sangiovanni-Vincentelli 1994).

An amplifier can be viewed in terms of its response to a DC input. An ideal inverting amplifier circuit would receive a DC input, invert it, and multiply it by the amplification factor. A circuit is flawed to the extent that it does not achieve the desired amplification; to the extent that the output signal is not centered on 0 volts (i.e., it has a bias); and to the extent that the DC response of the circuit is not linear.

We used a fitness measure based on SPICE's DC sweep. The DC sweep analysis measures the DC response of the circuit at several different DC input voltages. The circuits were analyzed with a 5 point DC sweep ranging from −10 millvolts (mv) to +10 mv, with input points at −10 mv, −5 mv, 0 mv, +5 mv, and +10 mv. SPICE then simulated the circuit's behavior for each of these five DC voltages. Four penalties (an amplification penalty, bias penalty, and two non-linearity penalties) are then derived.

First, the amplification factor of the circuit is measured by the slope of the straight line between the output for −10 mv and the output for +10 mv (i.e., between the outputs for the endpoints of the DC sweep). If the amplification factor is less than the target (60 dB), there is a penalty equal to the shortfall in amplification.

Second, the bias is computed using the DC output associated with a DC input of 0 volts. There is a penalty equal to the bias times a weight. A weight of 0.1 is used.

Third, the linearity is measured by the deviation between the slope of each of two shorter lines and the overall amplification factor of the circuit. The first shorter line segment connects the output value associated with an input of −10 mv and the output value for −5 mv. The second shorter line segment connects the output value for +5 mv and the output for +10 mv. There is a penalty for each of these shorter line segments equal to the absolute value of the difference in slope between the respective shorter line segment and the overall amplification factor of the circuit.

The fitness measure is multiobjective. Fitness is the sum of (1) the area of the bounding rectangle for the fully developed and laid-out circuit weighted by 10^{-6}, (2) the amplification penalty, (3) the bias penalty, and (4) the two non-linearity penalties; however, if this sum is less than 0.1 (indicating achievement of a very good amplifier), the fitness becomes simply the rectangle's area multiplied by 10^{-6}. Thus, after a good amplifier design is once achieved, fitness is based solely on area minimization.

Circuits that cannot be simulated by SPICE receive a penalty value of fitness (10^{8}).

3.5   Control Parameters

The population size, M, is 10,000,000. A maximum size of 300 points (functions and terminals) was established for each of the three result-producing branches for each program tree. The other control parameters are those that we have used on many other problems (Koza, Bennett, Andre, and Keane 1999, Appendix D).

3.6   Implementation on Parallel Computer

This problem was run on a home-built Beowulf-style (Sterling, Salmon, Becker, and Savarese 1999) parallel cluster computer system consisting of 1,000 350 MHz Pentium II processors (each accompanied by 64 megabytes of RAM). The system has a 350 MHz Pentium II computer as host. The processing nodes are connected with a 100 megabit-per-second Ethernet. The processing nodes and the host use the Linux operating system. The distributed genetic algorithm with unsynchronized generations and semi-isolated subpopulations was used with a subpopulation size of $Q = 10,000$ at each of $D = 1,000$ demes. As each processor (asynchronously) completes a generation, four boatloads of emigrants from each subpopulation are dispatched to each of the four toroidally adjacent processors. The 1,000 processors are hierarchically organized. There are $5 \times 5 = 25$ high-level groups (each containing 40 processors). If the adjacent node belongs to a different group, the migration rate is 2% and emigrants are selected based on fitness. If the adjacent node belongs to the same group, emigrants are selected randomly and the migration rate is 5% (10% if the adjacent node is in the same physical box).

4   Results

The best-of-generation circuit from generation 0 has a fitness of 999.86890.

The first best-of-generation circuit (figure 4) delivering 60 dB of amplification appears in generation 65. This 27-component circuit occupies an area of 8,234 and has an overall fitness of 33.042583.

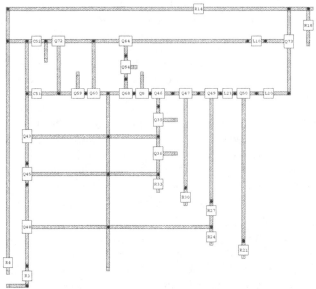

Figure 4 Best-of-run circuit from generation 65

The best-of-run circuit (figure 5) appears in generation 101. This circuit contains 11 transistors, 5 resistors, and 3 capacitors. The four "P" symbols indicate via's to the positive power supply. This 19-component circuit occupies an area of 4,751 and has an overall fitness of 0.004751. It occupies only 58% of the area of the 27-component circuit from generation 65. Note that figures 4 and 5 use different scales. Table 1 shows the number of components, the area, the four penalties comprising the non-area portion of the fitness measure, and the overall fitness for the these two circuits.

Table 1 Comparison of two best-of-generation circuits.

Generation	Components	Area	Four penalties	Fitness
65	27	8,234	33.034348	33.042583
101	19	4,751	0.061965	0.004751

The best-of-generation circuit from generations 65 has 81, 189, and 26 points, respectively, in its three branches. The best-of-run circuit from generation 101 has 65, 85, and 10 points, respectively, in its three branches. That is, the total size of both individuals and the size of each corresponding branch was reduced.

The third branches of these two individuals are both very small (26 and 10 points, respectively). The only effect of these branches are to insert a single transistor (a *pnp* transistor in the generation 65 and an *npn* transistor in generation 101).

The shaded portion of figure 4 shows the portion of the best circuit from generation 65 that is deleted in order to create the best circuit of generation 101.

The first branches of these two individuals are so similar that it is clear that these two branches are genealogically related. These two first branches account for 14 components (nine transistors and five resistors) that are in common with both circuits.

The second branches of these two individuals are almost completely different. The second branches account for the bulk of the reduction in component count (six transistors and three inductors) and the one added component (capacitor C10).

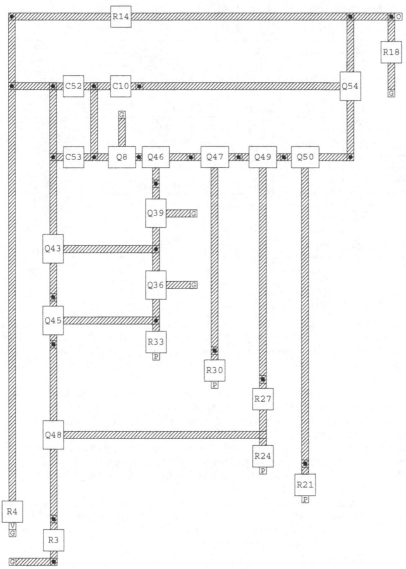

Figure 5 Best-of-run circuit from generation 101.

The difference between generations 65 and 101 caused by the first branches is that two extraneous components that are present in generation 65 are missing from the smaller 19-component circuit from generation 101.

Table 2 shows the *X* and *Y* coordinates for of the 19 components of the best circuit of generation 101 as well as the component value (sizing) for each capacitor and resistor and the type (*npn* q2n3904 or *pnp* q2n3904) for each transistor.

Table 2 Placement of the 19 components of the best circuit of generation 101.

Component	X coordinate	Y coordinate	Sizing / Type
Q8	-8.398678	21.184582	q2n3906
C10	-8.54565	31.121107	1.01e+02nf
R21	18.245857	-24.687471	1.48e+03k
R24	12.105233	-20.687471	5.78e+03k
R27	12.105233	-12.690355	3.61e+03k
R30	5.128666	-8.690355	8.75e+01k
R33	-3.398678	-4.690355	1.16e+03k
Q36	-3.398678	3.30961	q2n3906
Q39	-3.398678	13.309582	q2n3906
Q43	-18.472164	8.309597	q2n3904
Q45	-18.472164	-1.690355	q2n3904
Q46	-3.398678	21.184582	q2n3904
Q47	5.128666	21.184582	q2n3904
Q48	-18.472164	-17.687471	q2n3904
Q49	12.105233	21.184582	q2n3904
Q50	18.245857	21.184582	q2n3904
C52	-15.472164	31.121107	1.25e-01nf
C53	-15.472164	21.184582	7.78e+03nf
Q54	24.873787	31.121107	q2n3906

References

Bennett III, Forrest H, Koza, John R., Andre, David, and Keane, Martin A. 1996. Evolution of a 60 Decibel op amp using genetic programming. In Higuchi, Tetsuya, Iwata, Masaya, and Lui, Weixin (editors). *Proceedings of International Conference on Evolvable Systems: From Biology to Hardware (ICES-96)*. Lecture Notes in Computer Science, Volume 1259. Berlin: Springer-Verlag. Pages 455-469.

Garey, Michael R. and Johnson, David S. 1979. *Computers and Intractability: A Guide to the Theory of NP-Completeness*. New York, NY: W. H. Freeman.

Holland, John H. 1975. *Adaptation in Natural and Artificial Systems*. Ann Arbor, MI: University of Michigan Press.

Koza, John R., and Bennett III, Forrest H. 1999. Automatic synthesis, placement, and routing of electrical circuits by means of genetic programming. In Spector, Lee, Langdon, William B., O'Reilly, Una-May, and Angeline, Peter (editors). *Advances in Genetic Programming 3*. Cambridge, MA: MIT Press. Chapter 6. Pages 105 - 134.

Koza, John R., Bennett III, Forrest H, Andre, David, and Keane, Martin A. 1996. Automated design of both the topology and sizing of analog electrical circuits using genetic programming. In Gero, John S. and Sudweeks, Fay (editors). *Artificial Intelligence in Design '96*. Dordrecht: Kluwer Academic. Pages 151-170.

Koza, John R., Bennett III, Forrest H, Andre, David, and Keane, Martin A. 1999. *Genetic Programming III: Darwinian Invention and Problem Solving*. San Francisco, CA: Morgan Kaufmann.

Koza, John R., Bennett III, Forrest H, Andre, David, Keane, Martin A., and Brave Scott. 1999. *Genetic Programming III Videotape: Human-Competitive Machine Intelligence*. San Francisco, CA: Morgan Kaufmann.

Quarles, Thomas, Newton, A. R., Pederson, D. O., and Sangiovanni-Vincentelli, A. 1994. *SPICE 3 Version 3F5 User's Manual*. Department of Electrical Engineering and Computer Science, University of California, Berkeley, CA. March 1994.

Sterling, Thomas L., Salmon, John, and Becker, Donald J., and Savarese. 1999. *How to Build a Beowulf: A Guide to Implementation and Application of PC Clusters*. Cambridge, MA: The MIT Press.

Immunotronics : Hardware Fault Tolerance Inspired by the Immune System

D.W. Bradley* and A.M. Tyrrell

Department of Electronics, University of York,
Heslington, York, England.
dwb105,amt@ohm.york.ac.uk
http://www.amp.york.ac.uk/external/media/welcome.html

Abstract. An novel approach to hardware fault tolerance is proposed
that takes inspiration from the human immune system as a method
of fault detection and removal. The immune system has inspired work
within the areas of virus protection and pattern recognition yet its application to hardware fault tolerance is untouched. This paper introduces
many of the ingenious methods provided by the immune system to provide reliable operation and suggests how such concepts can inspire novel
methods of providing fault tolerance in the design of state machine hardware systems. Through a process of self/non-self recognition the proposed
hardware immune system will learn to differentiate between acceptable
and abnormal states and transitions within the 'immunised' system. Potential faults can then be flagged and suitable recovery methods invoked
to return the system to a safe state.

1 Introduction

The rapid increase in our understanding of the human body and genetics has
enabled us to gain insight into the miraculous operation of our own body. The
theory of evolution has been used within the field of evolutionary design, such as
genetic algorithms [15] and genetic programming [10]. Evolutionary theory has
also been applied to the subject of evolvable hardware to promote the design
of electronic systems without human intervention [14]. The last decade has seen
biological inspiration used as a source of fault tolerance. The development of the
body from a single cell through to a complete person has inspired the subject of
embryonics [4] [13] for development of fault tolerant systems.

The human body is a highly parallel architecture of asynchronous systems
and interacting processes that have been subject to evolution over millions of
years. Through the use of five external senses total cognition of the surrounding
environment is available. The human 'machine' also uses several distributed and
autonomous systems that maintain body stability and operation. The human
nervous system has inspired the most research to date with the development
of artificial neural networks (ANNs) that can be trained to learn and recognise

* Supported by the Engineering and Physical Sciences Research Council

J. Miller et al. (Eds.): ICES 2000, LNCS 1801, pp. 11–20, 2000.

features within a constrained environment. The immune system has also been a source of interest in the design of novel pattern recognition based applications including computer virus protection [9] and computer security [6]. An excellent review of immunity based systems can be found in [3].

This paper proposes that an immunological approach to the design and organisation of fault tolerant hardware may provide a novel alternative to the classical approaches of reliable system design - a subject that the authors have called *immunotronics* (immunological electronics), first introduced in [16].

Biologically inspired approaches to fault tolerance are discussed combined with an introduction to the key features of the immune system that make it excel in the task of fault tolerance. The subject of immunotronics is then introduced.

2   Biologically Inspired Fault Tolerance

Over thirty years ago, developments based on the challenge of designing reliable systems from unreliable components resulted in the notion of fault tolerance. Even after years of fault tolerant system research, the provision of dependable systems is still a very costly process limited to only the most critical of situations [1]. The typical approach is through the use of redundancy where functions are replicated by n versions of protected hardware. Embryonics is taking this a stage further through the cellular organisation and replication of hardware elements [4] [13].

The biological approach to fault tolerance is in the form of highly dependable distributed systems with a very high degree of redundancy. The human immune system protects the body from invaders, preventing the onset of chemical and cellular imbalances that may affect the reliable operation of the body. Similarities between the human immune system and the requirements of fault tolerant system design were first highlighted by Avizienis [1] who noted the potential analogies between hardware fault tolerance and the immune system. Use of the immune system as an approach to fault tolerance within systems was first noted and demonstrated in [17] for the design and operation of reliable software systems.

3   Key Immunological Features

Based upon the fundamental attributes of the immune system presented in [1], the following five key features can be noted:

- The immune system functions continuously and autonomously, only intervening the normal operation of the body when an invader or 'erroneous' condition is detected, much like in the presence of a faulty state. In a mapping to hardware, the analogy is that of fault detection and removal without the need for software support.
- The cells that provide the defence mechanisms are distributed throughout the body through its own communications network in the form of *lymphatic vessels* to serve all the organs. The hardware equivalent promotes distributed detection of faults with no centralised fault recognition and recovery.

- The immune cells that provide the detection mechanisms are present in large quantities and exist with a huge range of diversity. Limited diversity is already a common solution to fault tolerant system design.
- The immune system can learn and remember from past experiences what it should attack. The hardware analogy suggests the training (and possibly even continued improvement during operation) of fault detection mechanisms to differentiate between fault free and faulty states.
- Detection of invading antigens by the immune system is imperfect. The onset of faults in hardware systems is often due to the impossibility to exhaustively test a system. The testing phase can never test for every eventuality and so the analogy of imperfect detection suggests one remedy.

These five key features form the main aims of the immunotronic proposal. Section 4 now introduces how these are performed in the human immune system.

4   The Human Immune System

The human immune system (and that of all vertebrates) is unique amongst all living species in that it has evolved a complex genetic level defence mechanism for protection from invaders. This is implemented by the ability to detect the presence of foreign cells amongst those of the body, i.e. *self/non-self differentiation*.

4.1   Immune Architecture and Organisation

Defence against invaders, or *antigens* is accomplished through four distinguishable layers of protection [7] [8] from physical barriers, through physiological barriers in the forms of temperature and acidity to chemical and cellular interactions in the forms of innate and acquired immunity. Acquired immunity involves antibody and cell mediated immunity that defend against extra-cellular and intra-cellular infection respectively. The acquired immune system is the source of inspiration for this work.

The immune system cells are produced from stem cells in the bone marrow and divide to produce three types of cell:

- *Macrophages* are roaming scavenger cells that take part in both innate and acquired immunity. They perform a signalling role presenting fragments of antigens to other cells of the immune system.
- *B cells* can recognise antigens and produce a single type of *antibody* to counteract a specific antigen.
- *T cells* develop to form *helper*, *suppressor* and *killer* T cells. Helper and suppressor cells act as the master switched for the immune system by initiating and slowing down immune responses in the presence of an antigen. Killer T cells detect and destroy virus infected cells.

Response to an antigen occurs by the use of complementary receptors on the antigen and antibody known as *epitopes* and *paratopes* respectively. Both B and T cells have the ability to detect and counteract only one type of antigen and so huge diversity is a necessity. Such specificity means that at any one time there are over 10^{12} B cells within the body creating over 10^8 different types of antibody - a number impossible to encode for in the human genome of 10^5 genes. The rearrangement of antibody protein segments creates the huge variation needed. The presence of such a wide number of different antibodies means that an exact epitope-paratope match rarely occurs. An immune response can be initiated by approximate matching in a process called *affinity maturation* (figure 1). Under a continuous cycle of repeated optimisation, the B cells with the highest affinity to the invading antigen generate minor variants by *somatic hypermutation* resulting in an overwhelming quantity of antibodies to destroy the invading antigen.

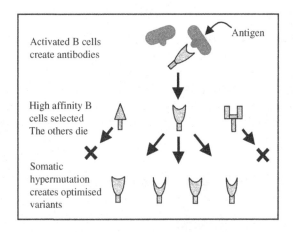

Fig. 1. Affinity maturation of antibodies

4.2   Antibody Mediated Immunity

Antibody mediated immunity protects the body from extra-cellular infection. B cells are constantly on patrol for antigens that they can bind to. If an approximate match occurs between a patrolling B cell and an antigen a response is initiated as in figure 2. Proliferation of antibodies only occurs if the corresponding T cells exists to stimulate the manufacture of optimised antibodies. When an antigen is encountered for the first time it takes several days for antibody proliferation to occur. Through the use of memory B and T cells secondary responses can provide a much more rapid response to the same infection at a later date.

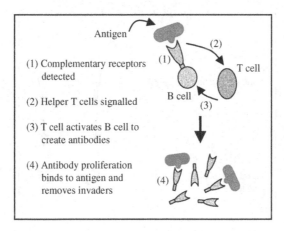

(1) Complementary receptors detected

(2) Helper T cells signalled

(3) T cell activates B cell to create antibodies

(4) Antibody proliferation binds to antigen and removes invaders

Fig. 2. Antibody mediated immunity

4.3 Self/Non-Self Discrimination

The process of antigen detection by random generation and complementary receptor matching is a very effective method of protection, but what prevents the immune system from binding to cell proteins that occur naturally within the body? Several theories have been proposed to explain how the immune system differentiates between self and non-self cells of the body [12]. The most widely accepted answer is that of *clonal deletion* and is demonstrated in figure 3. In contrast to the matured functional immune system with distributed censoring, a centralised development stage occurs first and carries out a process called *negative selection*. Immature helper T cells move to the *thymus* where self-proteins circulate through the thymus and are exposed to the helper T cells. If a maturing T cell binds to one of the self proteins it is destroyed. Only those T cells that are self tolerant survive to become fully functional activators of B cells.

5 The Immunotronic Proposal

The immunotronic proposal intents to explore the processes carried out by the human immune system to inspire new methods of fault tolerance that have already proven to be successful in nature. The likelihood of success is increased if relevant features of the immune system can be mapped to a form of hardware design. The notion of self and non-self is the major challenge in such a mapping. The approach taken is based upon the use of a finite state machine. The presence of valid states and transitions are compared to cells of the body, or *self*, and invalid states and transitions as antigens, or *non-self*. The proposals have used numerous analogies to the immune system. Xanthakis summarised the analogies for a software immune system in tabular form [17]. This has been used as a basis for the development of a similar feature mapping for the hardware fault

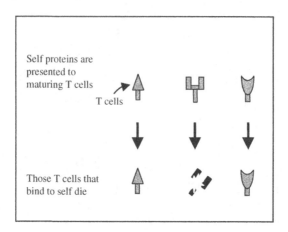

Fig. 3. Clonal deletion

tolerance. The mappings are summarised in tables 1 and 2 representing the entities, or physical elements of the system and then the set of processes that the immunotronic system may undergo. The tables are not exhaustive and present an early version of the potential mappings.

Immune System	Hardware Fault Tolerance
Self	Acceptable state/transition of states
Non-self (antigen)	Erroneous state/transition of states
Antibody (B cell)	Error tolerance conditions *(including epitope and paratope)*
Gene used to create antibody	Variables forming tolerance conditions
Paratope	Erroneous state/transition verification conditions
Epitope	Valid state/transition verification conditions
Helper T cell	Recovery procedure activator
Memory B/T cell	Sets of tolerance conditions *(epitope and paratope storage)*

Table 1. Entity feature mapping

5.1   The Learning Stage

The learning stage will correspond to the maturation of T cells in the thymus. The design and construction of any system will require a testing stage to validate the system before it is put into service. The results of this should provide sufficient data to give the system a sense of self. The current approach is to

Immune System	Hardware Fault Tolerance
Recognition of self	Recognition of valid state/transition
Recognition of non-self and malignant cells	Recognition of invalid state/transition
Learning during gestation	Learning of correct states and transitions
Antibody mediated immunity	Error detection and recovery
Antibody proliferation by clonal selection	Development of test vectors to improve the range of valid states and errors detected
Clonal deletion	Isolation of self-recognising tolerance conditions
Inactivation of antigen	Return to normal operation
Life of organism	Operation lifetime of the hardware

Table 2. Process feature mapping

'immunise' the state of the system by monitoring the inputs, current state and previous state of the system and storing each instance of self as a binary string. Although the human immune system does not store what is self, the fact that this information is already possessed makes it sensible to include. The nature of most systems means that they are too large to ensure full coverage of all valid conditions and states, so the importance of non-self and negative selection becomes relevant. Injection of faults into the system using typical fault models such as stuck-at-one, stuck-at-zero and tolerance checks should allow a sense of non-self to be defined. The major challenge of this work will be to ensure that the set of tolerance conditions created from the self and non-self data will be indicative of the entire system so if a condition is met that no match exists for within the immune system an assessment can be made of its validity. Potential solutions include the use of characteristic sub-string matching as used by Forrest [5] or an evolutionary approach to further tolerance condition generation. An evolutionary approach may even allow the system to continuously adapt and improve when in service.

5.2   Fault Recognition

In the recognition stage (figure 4) the states of the system can be compared to those of the stored T cells in memory and the validity of the current system state assessed. The memory is addressed and the validity of the current state returned. If the state is confirmed as valid then normal operation is allowed to continue. If a faulty state is detected a costimulation signal can activate any number of defined responses. A content addressable memory is suited to this process as all memory locations are searched in parallel. The importance of approximate matching must also be considered in this stage. The desire to provide only an indicative set of data means that complete tolerance strings may not always be available, it will be necessary to determine whether it is self, or non-self that the string matches, such as through a process of partial string matching.

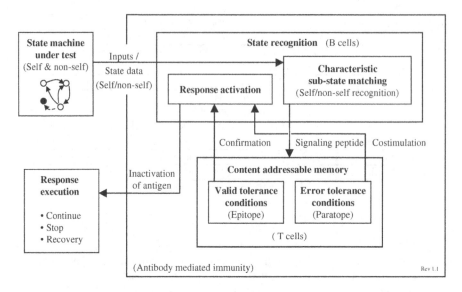

Fig. 4. The fault recognition stage

5.3   Recovery

The destruction or inactivation of a cell in the body is the ideal way of removing the invader due to the vast levels of redundancy. Such a process is rarely practical, although often necessary in hardware systems due to the finite level of redundancy available. In the presence of an intermittent error, a more ideal approach would be recovery or repair. The choice of recovery procedure could be a classical one that switches in a spare system. The immune inspired approach suggests potential recovery methods that ensure the system is returned to a default, or acceptable state as a form of forward error recovery minimising the need to switch in spare systems.

6   Results and Implementation Plans

Early work aims to concentrate on a centralised immunotronic approach to allow a system to be 'immunised' using many of the entity and processes mappings provided. Development of a state machine based counter with centralised immune inspired properties is currently underway. The platform consists of a Xilinx Virtex XCV300 FPGA development board from Virtual Computer Corporation [2]. The system currently performs simple recognition of valid bit strings made from the concatenation of the user inputs, previous and current state. An 'OK' signal flags the acceptance of the tolerance condition (figure 5). Stuck at faults can be injected to hold the state latches at either logic one or zero, irrespective of the internal logic of the state machine. This can then be used to assess the effects of missing tolerance conditions (figure 6). The next stage of this work will investigate the generation of an indicative set of both valid and invalid tolerance conditions. Future plans aim to advance this to a more decentralised system, such as that of multicellular hardware [13].

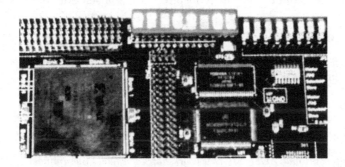

Fig. 5. The 'OK' signal confirms a state change from 7 to 8

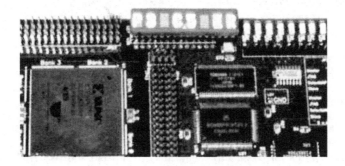

Fig. 6. The 'ER' signal flags the presence of an invalid transition from 3 to 5

7   Conclusion

The paper has introduced an immunological approach to hardware fault tolerance. It is believed that the methodology maps well to diverse system sizes where complete testing can often be a great challenge. The acquired immune response of the body is centrally learnt during growth through negative selection of self recognising T cells. This inspires similar approaches in the design of fault tolerant hardware to develop a sense of self and non-self. With the ever increasing

20   D.W. Bradley and A.M. Tyrrell

complexity in systems, fault avoidance is not a practical realisation. Faults must be tolerated and their effects minimised, just as the natural immune system does. The ideas proposed are somewhat different from the common approaches to fault tolerance but are nonetheless exciting and appealing if a realisable implementation can be achieved. Steps are now underway to develop the system and achieve the goal of biologically inspired self/non-self differentiation.

References

[1] A. Avizienis. Towards Systematic Design of Fault-Tolerant Systems. *IEEE Computer*, 30(4):51–58, April 1997.
[2] Virtual Computer Corporation. The Virtual Workbench, 1999.
http://www.vcc.com/VW.html.
[3] D. Dasgupta and N. Attoh-Okine. Immunity-Based Systems: A Survey. In *IEEE International Conference on Systems, Man and Cybernetics*, Orlando, 1997.
[4] S. Durand and C. Piguet. FPGA with Self-Repair Capabilities. In *FPGA '94, 2nd International ACM/SIGDA Workshop on Field-Programmable Gate Arrays*, pages 1–10, Berkeley, California, February 1994.
[5] S. Forrest et al. Self-Nonself Discrimination in a Computer. In *Proceedings of the 1994 IEEE Symposium on Research in Security and Privacy*, pages 202–212, Los Alamitos, CA, 1994. IEEE Computer Society Press.
[6] S. Forrest, S. Hofmeyr, and A. Somayaji. Computer Immunology. *Communications of the ACM*, 40(10):88–96, 1997.
[7] S. Hofmeyr. An Overview of the Immune System, 1997.
http://www.cs.umn.edu/~stevah/imm-html/introduction.html.
[8] C.A. Janeway and P. Travers. *Immunobiology, the Immune System in Health and Disease*. Churchill Livingstone, 2 edition, 1996.
[9] J.O. Kephart. A Biologically Inspired Immune System for Computers. In R.A. Brooks and P. Maes, editors, *Artificial Life IV, Proceedings of the Fourth International Workshop on the Synthesis and Simulation of Living Systems*, pages 130–139. MIT Press, 1994.
[10] J.R. Koza. *Genetic Programming*. MIT Press, 1992.
[11] P. Marchal, C. Piguet, D. Mange, G. Tempesti, and S. Durand. BioBIST Biology and Built-In Self-Test Applied to Programmable Architectures. In *IEEE Built in Self Test/Design for Testability Workshops*, Vail, Colorado (USA), April 1994.
[12] P. Marrack and J.W. Kappler. How the Immune System Recognises the Body. *Scientific American*, pages 49–55, September 1993.
[13] C. Ortega-Sánchez and A.M. Tyrrell. Design of a Basic Cell to Construct Embryonic Arrays. In *IEE Proceedings on Computers and Digital Techniques*, volume 143, pages 242–248, May 1998.
[14] A. Thompson. Silicon evolution. In *Genetic Programming 1996 : Proceedings of the First Annual Conference*, pages 444–452, 1996.
[15] M. Tomassini. A Survey of Genetic Algorithms. *Annual Reviews of Computational Physics*, 3:87–118, 1995.
[16] A.M. Tyrrell. Computer Know Thy Self! : A Biological Way to look af Fault Tolerance. In *2nd Euromicro/IEEE Workshop on Dependable Computing Systems*, Milan, 1999.
[17] S. Xanthakis, S. Karapoulios, R. Pajot, and A. Rozz. Immune System and Fault Tolerant Computing. In J.M. Alliot, editor, *Artificial Evolution*, volume 1063 of *Lecture Notes in Computer Science*, pages 181–197. Springer-Verlag, 1996.

Ant Colony System for the Design of Combinational Logic Circuits

Carlos A. Coello Coello†, Rosa Laura Zavala G.‡, Benito Mendoza García‡,
and Arturo Hernández Aguirre§

†Laboratorio Nacional de Informática Avanzada
Rébsamen 80, A.P. 696
Xalapa, Veracruz, México 91090
ccoello@xalapa.lania.mx
‡MIA, LANIA-UV
Sebastián Camacho 5
Xalapa, Veracruz, México
{rzavala,bmendoza}@mia.uv.mx
§EECS Department
Tulane University
New Orleans, LA 70118, USA
hernanda@eecs.tulane.edu

Abstract. In this paper we propose an application of the Ant System
(AS) to optimize combinational logic circuits at the gate level. We define
a measure of quality improvement in partially built circuits to compute
the distances required by the AS and we consider as optimal those solutions that represent functional circuits with a minimum amount of gates.
The proposed methodology is described together with some examples
taken from the literature that illustrate the feasibility of the approach.

1   Introduction

In this paper we extend previous work on the optimization of combinational
logic circuits [1, 2], by experimenting with a metaheuristic: the ant system (AS)
[6, 3].

The AS is a multi-agent system where low level interactions between single
agents (i.e., artificial ants) result in a complex behavior of the whole ant colony.
The idea was inspired by colonies of real ants, which deposit a chemical substance
on the ground called *pheromone* [5]. This substance influences the behavior of
the ants: they will tend to take those paths where there is a larger amount of
pheromone. The AS was originally proposed for the traveling salesman problem
(TSP), and according to Dorigo [6], to apply efficiently the AS, it is necessary to
reformulate our problem as one in which we want to find the optimal path of a
graph and to identify a way to measure the distances between nodes. This might
not be an easy or obvious task in certain applications like the one presented
in this paper. Therefore, we will provide with a detailed discussion of how to
reformulate the circuit optimization problem as to allow the use of the AS, and
we will present several examples to illustrate the proposed approach.

J. Miller et al. (Eds.): ICES 2000, LNCS 1801, pp. 21–30, 2000.

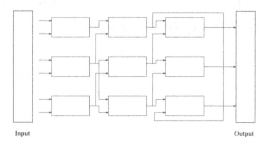

Fig. 1. Matrix used to represent a circuit to be processed by an agent (i.e., an ant). Each gate gets its inputs from either of the gates in the previous column.

2   Description of the Aproach

Since we need to view the circuit optimization problem as one in which we want to find the optimal path of a graph, we will use a matrix representation for the circuit as shown in Fig. 1. This matrix is encoded as a fixed-length string of integers from 0 to $N - 1$, where N refers to the number of rows allowed in the matrix.

More formally, we can say that any circuit can be represented as a bidimensional array of gates $S_{i,j}$, where j indicates the *level* of a gate, so that those gates closer to the inputs have lower values of j. (Level values are incremented from left to right in Fig. 1). For a fixed j, the index i varies with respect to the gates that are "next" to each other in the circuit, but without being necessarily connected. Each matrix element is a gate (there are 5 types of gates: AND, NOT, OR, XOR and WIRE) that receives its 2 inputs from any gate at the previous column as shown in Fig. 1.

We have used this representation before with a genetic algorithm (GA) [1, 2]. The path of an agent (i.e., an ant) will then be defined as the sub-portion of this matrix (of a certain pre-defined maximum size) representing a Boolean expression (i.e., an ant will build a circuit while traversing a path). Each state within the path is a matrix position of the circuit and the distance between two states of the path is given by the increase or decrease of the cost of the circuit when moving from one state to the following. Such cost is defined in our case in terms of the number of gates used—i.e., a feasible circuit that uses the minimum amount of gates possible is considered optimal. The aim is to maximize a certain payoff function. Since our code was built upon our previous GA implementation, we adopted the use of fixed matrix sizes for all the agents, but this need not be the case (in fact, we could represent the Boolean expressions directly rather than using a matrix). The matrix containing the solution to the problem is built in a column-order fashion following the steps described next.

The gate and inputs to be used for each element of the matrix are chosen randomly from the set of possible gates and inputs (a modulo function is used when the relationship between inputs and matrix rows is not one-to-one). Each

Input 1	Input 2	Gate Type

Fig. 2. Encoding used for each of the matrix elements that represent a circuit.

state is, therefore, a triplet in which the first 2 elements refer to each of the inputs used (taken from the previous level or column of the matrix) and the third is the corresponding gate (chosen from AND, OR, NOT, XOR, WIRE (WIRE basically indicates a null operation, or in other words, the absence of gate) as shown in Fig. 2 (only 2-input gates were used in this work). For the gates at the first level (or column), the possible inputs for each gate were those defined by the truth table given by the user (a modulo function was implemented to allow more rows than available inputs).

One important difference between the statement of this problem and the TSP is that in our case not all the states within the path have to be visited, but both problems share the property that the same state is not to be visited more than once (this property is also present in some routing applications [4]).

When we move to another state in the path, a value is assigned to all the states that have not been visited yet and the next state (i.e., the next triplet) is randomly selected using a certain selection factor p. This selection factor determines the chance of going from state i to state j at the iteration t, and is computed using the following formula that combines the pheromone trail with the heuristic information used by the algorithm:

$$p_{i,j}^k(t) = f_{i,j}(t) \times h_{i,j} \tag{1}$$

where k refers to the ant whose pheromone we are evaluating, t refers to the current iteration, $f_{i,j}(t)$ is the amount of pheromone between state i and state j, and $h_{i,j}$ is the score increment between state i and state j. This score is measured according to the number of matches between the output produced by the current circuit and the output desired according to the truth table given by the user. This score increment ($h_{i,j}$) is analogous to the distance between nodes used in the TSP. No normalization takes place at this stage, because in a further step of the algorithm a proportional selection process is performed.

The amount of pheromone is updated each time an agent builds an entire path (i.e., once the whole circuit is built). Before this update, the pheromone evaporation is simulated using the following formula:

$$f_{i,j}(t+1) = (1 - \alpha) \times f_{i,j}(t) + \sum_{k=1}^{m} f_{i,j}^k(t) \tag{2}$$

where $0 < \alpha < 1$ ($\alpha = 0.5$ was used in all the experiments reported in this paper) is the trail persistence and its use avoids the unlimited accumulation of pheromone in any path, m refers to the number of agents (or ants) and $\sum_{k=1}^{m} f_{i,j}^k(t)$ corresponds to the total amount of pheromone deposited by all the ants that went through states (i, j). Furthermore, the pheromone trail is updated according to the circuit built by each agent. The pheromone of the gates of the first row of each column is increased using the following criteria:

1) If the circuit is not feasible (i.e., if not all of its outputs match the truth table), then:

$$f_{i,j}^k = f_{i,j}^k + \text{payoff} \tag{3}$$

2) If the circuit is feasible (i.e., all of its outputs match the truth table), then:

$$f_{i,j}^k = f_{i,j}^k + (\text{payoff} \times 2) \tag{4}$$

3) The best individual (from all the agents considered) gets a larger reward:

$$f_{i,j}^k = f_{i,j}^k + (\text{payoff} \times 3) \tag{5}$$

The value of "payoff" is given by the number of matches produced between the output generated by the circuit built by the agent and the truth table given by the user (a bonus is added for each WIRE found in the solution only in those cases in which the circuit is feasible—i.e., it matches all the outputs given in the truth table).

To build a circuit, we start by placing a gate (randomly chosen) at a certain matrix position and we fill up the rest of the matrix using WIREs. This tries to compute the effect produced by a gate used at a certain position (we compute the score corresponding to any partially built circuit). The distance is computed by subtracting the hits obtained at the current level (with respect to the truth table) minus the hits obtained up to the previous level (or column). When we are at the first level, we assume a value of zero for the previous level.

Table 1. Truth table for the circuit of the first example.

X	Y	Z	F
0	0	0	0
0	0	1	0
0	1	0	0
0	1	1	1
1	0	0	0
1	0	1	1
1	1	0	1
1	1	1	0

3   Results

We used several examples taken from the literature to test our AS implementation. Our results were compared to those obtained by two human designers and a genetic algorithm with binary representation (see [2] for details).

3.1   Example 1

Table 2. Comparison of results between the AS, a binary GA (BGA), and two human designers for the circuit of the first example.

BGA	Human Designer 1
$F = Z(X + Y) \oplus (XY)$	$F = Z(X \oplus Y) + Y(X \oplus Z)$
4 gates	5 gates
2 ANDs, 1 OR, 1 XOR	2 ANDs, 1 OR, 2 XORs
AS	**Human Designer 2**
$F = (Z \oplus XY)(X + Y)$	$F = X'YZ + X(Y \oplus Z)$
4 gates	6 gates
2 ANDs, 1 OR, 1 XOR	3 ANDs, 1 OR, 1 XOR, 1 NOT

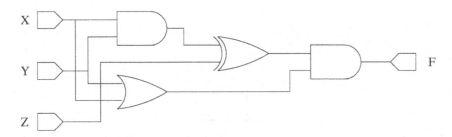

Fig. 3. Circuit produced by the AS for the first example.

Our first example has 3 inputs and one output, as shown in Table 1. In this case, the matrix used was of size 5×5, and the length of each string representing a circuit was 75. Since 5 gates were allowed in each matrix position, then the size of the intrinsic search space (i.e., the maximum size allowed as a consequence of the representation used) for this problem is 5^l, where l refers to the length required to represent a circuit ($l = 75$ in our case). Thefore, the size of the intrinsic search space is $5^{75} \approx 2.6 \times 10^{52}$. The graphical representation of the circuit produced by the AS is shown in Fig. 3. The AS found this solution after 13 iterations using 30 ants.

The comparison of the results produced by the AS, a genetic algorithm with binary representation (BGA) and two human designers are shown in Table 2. As we can see, the AS found a solution with the same number of gates as the BGA. In this case, human designer 1 used Karnaugh Maps plus Boolean algebra identities to simplify the circuit, whereas human designer 2 used the Quine-McCluskey Procedure.

The parameters used by the BGA were the following: crossover rate = 0.5, mutation rate = 0.0022, population size = 900, maximum number of generations

= 400. The solution reported for the BGA in Table 2 was found in generation 197. The matrix used by the BGA was of size 5 × 5.

3.2   Example 2

Table 3. Truth table for the circuit of the second example.

Z	W	X	Y	F
0	0	0	0	1
0	0	0	1	1
0	0	1	0	0
0	0	1	1	1
0	1	0	0	0
0	1	0	1	0
0	1	1	0	1
0	1	1	1	1
1	0	0	0	1
1	0	0	1	0
1	0	1	0	1
1	0	1	1	0
1	1	0	0	0
1	1	0	1	1
1	1	1	0	0
1	1	1	1	0

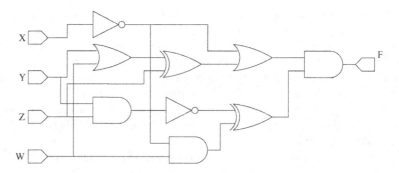

Fig. 4. Circuit produced by the AS for the second example.

Our second example has 4 inputs and one output, as shown in Table 3. In this case, the matrix used was of size 10 × 8. The size of the intrinsic search space for this problem is then 5^{240}. The graphical representation of the circuit

Table 4. Comparison of results between the AS, a binary GA (BGA), a human designer and Sasao's approach for the circuit of the second example

BGA
$F = (WYX' \oplus ((W + Y) \oplus Z \oplus (X + Y + Z)))'$
10 gates
2 ANDs, 3 ORs, 3 XORs, 2 NOTs
Human Designer 1
$F = ((Z'X) \oplus (Y'W')) + ((X'Y)(Z \oplus W'))$
11 gates
4 ANDs, 1 OR, 2 XORs, 4 NOTs
AS
$F = (((W + Y) \oplus Z) + X')((YZ)' \oplus (X'W))$
9 gates
3 ANDs, 2 ORs, 2 XORs, 2 NOTs
Sasao
$F = X' \oplus Y'W' \oplus XY'Z' \oplus X'Y'W$
12 gates
3 XORs, 5 ANDs, 4 NOTs

produced by the AS is shown in Fig. 4. The AS found this solution after 15 iterations using 30 ants.

The comparison of the results produced by the AS, a genetic algorithm with binary representation (BGA), a human designer (using Karnaugh maps), and Sasao's approach [7] are shown in Table 4. In this case, the AS found a solution slightly better than the BGA. Sasao has used this circuit to illustrate his circuit simplification technique based on the use of ANDs & XORs. His solution uses, however, more gates than the circuit produced by our approach.

The parameters used by the BGA for this example were the following: crossover rate = 0.5, mutation rate = 0.0022, population size = 2000, maximum number of generations = 400. Convergence to the solution shown for the BGA in Table 4 was achieved in generation 328. The matrix used by the BGA was of size 5×5.

3.3   Example 3

Our third example has 4 inputs and one output, as shown in Table 5. In this case, the matrix used was of size 15×15. The size of the intrinsic search space for this problem is then 5^{675}. The graphical representation of the circuit produced by the AS is shown in Fig. 5. The AS found this solution after 8 iterations using 30 ants.

The comparison of the results produced by the AS, a genetic algorithm with binary representation (BGA), and 2 human designers (the first using Karnaugh maps and the second using the Quine-McCluskey procedure), are shown in Table 6. In this example, the BGA found a solution slightly better than the AS.

Table 5. Truth table for the circuit of the third example.

A	B	C	D	F
0	0	0	0	1
0	0	0	1	0
0	0	1	0	0
0	0	1	1	0
0	1	0	0	1
0	1	0	1	1
0	1	1	0	1
0	1	1	1	1
1	0	0	0	1
1	0	0	1	1
1	0	1	0	1
1	0	1	1	0
1	1	0	0	0
1	1	0	1	1
1	1	1	0	0
1	1	1	1	1

Table 6. Comparison of results between the AS, a binary GA (BGA), and two human designers for the circuit of the third example.

BGA
$F = ((A \oplus B) \oplus AD) + (C + (A \oplus D))'$
7 gates
1 AND, 2 ORs, 3 XORs, 1 NOT

Human Designer 1
$F = ((A \oplus B) \oplus ((AD)(B + C))) + ((A + C) + D)'$
9 gates
2 ANDs, 4 ORs, 2 XORs, 1 NOT

AS
$F = ((B \oplus D) \oplus (A + D)) \oplus ((B + C) + (A \oplus D))'$
8 gates
3 ORs, 4 XORs, 1 NOT

Human Designer 2
$F = A'B + A(B'D' + C'D)$
10 gates
4 ANDs, 2 ORs, 4 NOTs

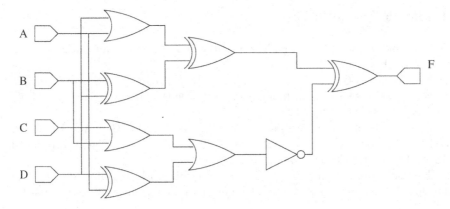

Fig. 5. Circuit produced by the AS for the third example.

The parameters used by the BGA for this example were the following: crossover rate = 0.5, mutation rate = 0.0022, population size = 2600, maximum number of generations = 400. Convergence to the solution shown for the BGA in Table 6 was achieved in generation 124. The matrix used by the BGA was of size 5×5.

4   Conclusions and Future Work

In this paper we have presented an approach to use the ant colony system to optimize combinational logic circuits (at the gate level). The proposed approach was described and a few examples of its use were presented. Results compared fairly well with those produced with a BGA (a GA with binary representation) and are better than those obtained using Karnaugh maps and the Quine-McCluskey Procedure.

Some of the future research paths that we want to explore are the parallelization of the algorithm to improve its performance (each agent can operate independently from the others until they finish a path and then they have to be merged to update the pheromone trails) and the hybridization with other algorithms (e.g., local search).

We also want to experiment with other metaheuristics such as tabu search to scale up the use of AS to larger circuits (our current implementation is limited to circuits of only one output) without a significant performance degradation. Finally, we are also interested in exploring alternative (and more powerful) representations of a Boolean expression in an attempt to overcome the inherent limitations of the matrix representation currently used to solve real-world circuits in a reasonable amount of time and without the need of excessive computer power.

Acknowledgements

The authors would like to thank the anonymous reviewers for their valuable comments that helped them improve this paper.

We also thank Carlos E. Mariano for supplying the initial code for the AS on which this work is based. The first author acknowledges support from CONACyT through project No. I-29870 A.

The second and third authors acknowledge support from CONACyT through a scholarship to pursue graduate studies at the Maestría en Inteligencia Artificial of LANIA and the Universidad Veracruzana.

The last author acknowledges support for this work to DoD EPSCoR and the Board of Regents of the State of Louisiana under grant F49620-98-1-0351.

References

1. Carlos A. Coello, Alan D. Christiansen, and Arturo Hernández Aguirre. Automated Design of Combinational Logic Circuits using Genetic Algorithms. In D. G. Smith, N. C. Steele, and R. F. Albrecht, editors, *Proceedings of the International Conference on Artificial Neural Nets and Genetic Algorithms*, pages 335–338. Springer-Verlag, University of East Anglia, England, April 1997.
2. Carlos A. Coello, Alan D. Christiansen, and Arturo Hernández Aguirre. Use of Evolutionary Techniques to Automate the Design of Combinational Circuits. *International Journal of Smart Engineering System Design*, 1999. (accepted for publication).
3. A. Colorni, M. Dorigo, and V. Maniezzo. Distributed optimization by ant colonies. In F. J. Varela and P. Bourgine, editors, *Proceedings of the First European Conference on Artificial Life*, pages 134–142. MIT Press, Cambridge, MA, 1992.
4. G. Di Caro and M. Dorigo. AntNet: Distributed Stigmergetic Control for Communications Networks. *Journal of Artificial Intelligence Research*, 9:317–365, 1998.
5. M. Dorigo and G. Di Caro. The Ant Colony Optimization Meta-Heuristic. In D. Corne, M. Dorigo, and F. Glover, editors, *New Ideas in Optimization*. McGraw-Hill, 1999.
6. M. Dorigo, V. Maniezzo, and A. Colorni. Positive feedback as a search strategy. Technical Report 91-016, Dipartimento di Elettronica, Politecnico di Milano, Italy, 1991.
7. Tsutomu Sasao, editor. *Logic Synthesis and Optimization*. Kluwer Academic Press, 1993.

Evolving Cellular Automata for Self-Testing Hardware

Fulvio Corno, Matteo Sonza Reorda, Giovanni Squillero

Politecnico di Torino
Dipartimento di Automatica e Informatica
Corso Duca degli Abruzzi 24 I-10129, Torino, Italy
{corno, sonza, squillero}@polito.it

Abstract. Testing is a key issue in the design and production of digital circuits: the adoption of BIST (Built-In Self-Test) techniques is increasingly popular, but requires efficient algorithms for the generation of the logic which generates the test vectors applied to the Unit Under Test. This paper addresses the issue of identifying a Cellular Automaton able to generate input patterns to detect stuck-at faults inside a Finite State Machine (FSM) circuit. Previous results already proposed a solution based on a Genetic Algorithm which directly identifies a Cellular Automaton able to reach good Fault Coverage of the stuck-at faults. However, such method requires 2-bit cells in the Cellular Automaton, thus resulting in a high area overhead. This paper presents a new solution, with an area occupation limited to 1 bit per cell; the improved results are possible due to the adoption of a new optimization algorithm, the *Selfish Gene* algorithm. Experimental results are provided, which show that in most of the standard benchmark circuits the Cellular Automaton selected by the Selfish Gene algorithm is able to reach a Fault Coverage higher that what can be obtained with current engineering practice with comparable area occupation.

1 Introduction

Built-In Self-Test (BIST) [ABFr90] has been widely recognized as an effective approach for testing of Application Specific Integrated Circuits (ASICs). In the last decade, successful adoption of BIST has been reported for circuits as a whole and for embedded macros. In the meantime, design techniques evolved significantly, and automatic synthesis tools are now commonly used, especially for Finite State Machine (FSM) synthesis. Deeply embedded, automatically synthesized FSMs acting as Control Units can often be found in current designs, and resorting to BIST is an attracting approach for their test. Fig. 1 shows the structure of a BIST circuit (or macro): during the normal working mode, the *Unit Under Test* (UUT) is fed with the values coming from the circuit Primary Inputs and its outputs drive the circuit Primary Outputs. When the Test Mode is selected through the Normal/Test signal, the UUT is fed by a special circuitry named *Input Pattern Generator* producing the vectors which activate the possible faults inside the UUT. The *Output Data Evaluator* checks whether the output behavior of the UUT matches the expected values. The *BIST Controller* manages the whole test circuitry and possibly generates the Good/Faulty signal to the outside.

J. Miller et al. (Eds.): ICES 2000, LNCS 1801, pp. 31–40, 2000.

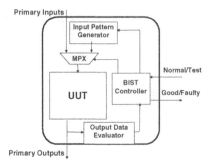

Figure 1: Architecture of a BIST circuit.

One of the today main issues in the test area is how to exploit BIST circuits to test embedded FSMs, facing the problems coming from their limited accessibility and their highly sequential behavior.

Several solutions have been proposed: partial and full scan [Jett95], CSTP [KrPi89], BILBO [KMZw79]. Most of the approaches tend to transform circuits into combinational ones for the purpose of test. Their effectiveness can be evaluated from many points of view: how much area and performance overhead they introduce, which fault coverage they guarantee, how much delay they introduce in the critical path, how easily and automatically they can be introduced into the original design structures, etc.

Any BIST solution includes a mechanism for generating test patterns, and another for evaluating the output behavior. When FSMs are not transformed into combinational circuits during test, the former issue is by far the most critical one, as many faults can be detected only provided that specific sequences are applied to the inputs. These sequences can be generated by Automatic Test Pattern Generators, but hardware structures able to reproduce them (e.g., based on ROMs) are very expensive from the area and performance points of view. Cellular Automata (CA) have already been proposed as random input pattern generators [HMPM89] and for reproducing deterministic unordered input vectors [BoKa95] [SCDM94] for tsting combinatorial circuits. Previous attempts to exploit CA to reproduce deterministic *ordered* input vectors [BoKa95], that are necessary for testing sequential circuits, limited themselves to prove the difficulty of attaining any useful result. Experimental results we gathered confirm that it is very difficult to identify any CA able to reproduce a given sequence, when this is longer than some tens of vectors.

The problem was solved in the past [CPSo96] by means of a Genetic Algorithm that computed the optimal set of CA rules by interacting with a fault simulator to evaluate the fitness of potential solutions. This approach was shown to generate satisfactory results in terms of fault coverage [CCPS97], but the area overhead was higher than competing approaches. The main reason behind excessive area occupation was the necessity of using CA cells with 4 possible states, that required 2 flip-flops each. An attempt to reduce this area was presented in [CGPS98], where the CA generating sequences was re-used also for analyzing the circuit output values, but for some classes of circuits this approach causes a decrease in the attained fault coverage.

This paper presents a new approach to the identification of a set of CA rules, that is able to achieve satisfactory fault coverage results with a reduced area overhead. In particular, we adopt the Selfish Gene algorithm [CSSq99a] instead of a more traditional GA, and this allows us to obtain good results even with 1-bit cells, thus effectively cutting in half area occupation with respect to previous techniques. Experimental results show that the attained fault coverage is substantially higher than what can be obtained by any previously proposed method with comparable area requirements, including a maximum-length CA [HMPM89] and that described in [CCPS97].

Section 2 reports some basics about CA and describes the hardware structure we adopted. Section 3 describes the Selfish Gene algorithm we adopted for the automatic synthesis of the Cellular Automaton starting from the netlist of the addressed circuit. Section 4 reports some preliminary experimental results, and Section 5 draws some conclusions.

2   Hardware architecture

2.1   Cellular Automata

Due to their versatility and ease of reconfiguration, in this paper we investigate the use of Cellular Automata for implementing IPG blocks.

A cellular automaton [Wolf83] [ToMa87] is a system composed of cells, whose behavior advances in time in discrete steps. Cells are connected in regular structures (*grids*), where the cells directly connected to a given cell are referred to as its *neighbors*. A *state* is associated to each cell. Each cell communicates its present state to its neighbors and computes its new state from its current state and from that of its neighbors.

The new state computation law is given by a *rule* characterizing each cell in the system. In the case of *binary* Cellular Automata, the state of each cell can be either 0 or 1: in this case the evolution rule is best expressed as a truth table, which lists the new state of the cell as a Boolean function. These truth tables are often expressed as decimal numbers, by interpreting them as a binary code.

The behavior of a CA is therefore specified by giving:
- the *structure* of the interconnection grid. The most used are one-, two- or three-dimensional square grids, but hyper-cubes, triangulated meshes, and trees can also be found.
- the *neighborhood* of each cell. Given a grid, the neighborhood is usually defined as the set of unit distance cells, where the distance is usually counted as the number of *hops* over the grid.
- the *boundary conditions*. Whenever the grid is not infinite, boundary conditions specify how the neighborhood of boundary cells (that would otherwise be incomplete) should be constructed. Either null boundary conditions (i.e., assuming that the grid is surrounded by cells whose state is invariably zero) or cyclic

boundary conditions (i.e., assuming that the CA grid is circular, and that cells lying on opposite boundaries of the grid are adjacent) are usually adopted.

- the *evolution rules* for each cell in the grid. A CA is defined as *uniform* when all of its cells follow the same evolution rule. More often, different cells follow different rules, and in this case the CA is said to be *non uniform* or *hybrid* (HCA).
- the *initial conditions*. It is usually assumed that the evolution of the CA starts from some predetermined non-null configuration.

2.2   Cellular Automata for BIST

When adopting Hybrid Cellular Automata (HCA) for BIST circuits, a standard test-per-clock architecture (Figure 2) is usually adopted, where the CA feeds the inputs of the Uircuit Under Test (UUT), and an Output Data Evaluator (ODE) monitors the Primary Outputs.

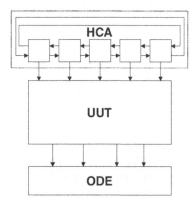

Figure 2: Test-per-clock architecture

In this paper, a one-dimensional CA with one cell for each circuit input is adopted. Each cell has two neighbors, therefore its next state is determined as a Boolean function of 3 bits: the present state of the cell and that of its neighbors. The rule of each cell is therefore selected among a set of $2^{(2^3)}=256$ different Boolean functions.

While in previous works we adopted a CA architecture with two flip-flops per cell, we are now able to obtain good results even with one flip-flop per cell. The chosen architecture is a one-dimensional, hybrid CA with 2 states (0 and 1) per cell and cyclic boundary conditions. The adopted CA has as many cells as the number of circuit PIs. Each cell S_i is composed (Fig. 3) of a combinational network f_i (the cell rule) fed with the outputs of the two adjacent cells S_{i-1} and S_{i+1} (the cell neighborhood) and with the value stored in S_i. The output of f_i is connected to the data input of the cell flip-flop.

With this choices, for an n-input circuit, n flip-flops and n combinational blocks are used.

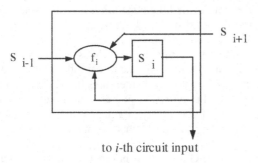

to *i*-th circuit input

Figure 3: Architecture of the *i*-th stage.

Due to the easier mathematical model, *linear* HCA configurations are often used by hardware designers as IPG. One frequently used configuration is the so-called 90/150 CA [CaZa95], that is composed of cells programmed with either rule 90 or rule 150. Rules 90 and 150 are composed exclusively of an ex-or gate, and can be combined to achieve maximum-length pseudo-random sequences as shown in [CaZa95].

Although this configuration, that generates pseudo-random test patterns, is effective from the point of view of fault coverage of *combinational* circuits, it is not optimized for testing *sequential* ones, where random vectors are seldom able to visit all the circuit states. In order to deal with this additional constraint, we had to exploit new degrees of freedom, by moving into the field of *non-linear* HCA, where f_i is not composed by ex-or gates, only.

We start with a 90/150 linear HCA, and improve it with an Evolutionary Algorithm until the fault coverage reaches acceptable levels. This approach can be implemented resorting to the SG algorithm detailed below, since it allows starting with the selected 90/150 linear HCA as initial solution, and benefits as a sort of intelligent local search.

3   The algorithm

3.1   The Selfish Gene Algorithm

The *Selfish Gene algorithm* (SG) is an evolutionary optimization algorithm based on a recent interpretation of the Darwinian theory. It evolves a population of individuals seeking for the fittest one. In the *selfish gene* biological theory, population itself can be simply seen as a *pool of genes* where the number of individuals, and their specific identity, are not of interest. Therefore, differently from other evolutionary algorithms, the SG resorts to a statistical characterization of the population, by representing and evolving some statistical parameters only. Evolution proceeds in discrete steps: individuals are extracted from the population, collated in tournaments and winner offspring is allowed to spread back into the population.

An individual is identified by the list of its genes. The whole list of genes is called *genome* and a position in the genome is termed *locus*. Each locus can be occupied by different genes. All these candidates are called the gene *alleles*. In the context of an optimization problem, looking for the fittest individual corresponds to determine the best set of genes according to the function to be optimized.

Since the SG "virtual" population is unlimited, individuals can be considered unique, but some genes would certainly be more frequent than others might. At the end of the evolution process, the *frequency* of a gene measures its *success* against its alleles. However, at the beginning of the evolution process, the frequency can be regarded as the gene *desirability*. When the majority of a population is characterized by the presence of a certain characteristic, new traits must harmonize with it in order to spread.

The pseudo-code of the core of the SG algorithm is reported in Fig. 4. Further implementation details about the SG algorithm are available in [CSSq98a], while biological motivations are better analyzed in [CSSq98b]. An extension to the algorithm to deal with more complex fitness landscapes is in [CSSq99].

```
genome SG(VirtualPopulation P)
{
        genome BEST, G₁, G₂, winner, loser;

        iter = 0;
        BEST = select_individual(P);      // best so far
        do {
           ++iter;
           G₁ = select_individual(P);
           G₂ = select_individual(P);
           tournament(G₁, G₂); // identify winner and loser
           increase_allele_frequencies(P, winner);
           decrease_allele_frequencies(P, loser);
           if(winner is preferable to BEST)
                BEST = winner;
        } while(steady_state()==FALSE && iter<max_iter) ;
        return BEST;
}
```

Figure 4: Selfish Gene algorithm pseudo-code.

3.2   Computation of optimal CA rules

In our approach, the SG algorithm is exploited to search in the space of $2^{(2^3)}$ rules that can be used for each cell, looking for the CA which maximizes the attained Fault Coverage.

Let us consider a gate-level description of a circuit. In our BIST approach, we assume that the Input Pattern Generator is an n-cell CA whose outputs are connected to the n circuit inputs. Our goal is to identify a CA able to generate an input sequence detecting the highest number of faults. By detecting a fault, we mean that the output values produced by the good and faulty circuits differ during at least one clock period. We do not discuss here how to analyze the circuit outputs. The permanent single stuck-at fault model is adopted.

Our algorithm reads the circuit netlist and the fault list, and chooses a rule for each CA cell: it is also able to provide the number of faults detected by the input sequence generated by this structure when it runs for a given number of T clock cycles.

In the Selfish Gene algorithm, each chromosome corresponds to a CA: the value of the i-th gene identifies the rule for the i-th stage of the CA. As described in the previous Section, the adopted CA structure allows for $2^{(2^3)}$ different rules for each stage: each gene can assume any value between 0 and $2^{(2^3)}$-1. The initial population is initialized by strongly biasing towards the canonical 90/150 solution, in order to reduce CPU times by starting from an already good solution. The Selfish Gene algorithm is then asked to improve that solution.

The evaluation function associated to each chromosome is the Fault Coverage attained by the sequence generated by the corresponding CA when run for T clock cycles. The evaluation function is computed by generating the sequence and then fault simulating it. Efficient fault simulation techniques [NCPa92] have been exploited to implement the procedures performing this task. The initial state of the CA is the all-0s state.

4   Experimental Results

We implemented the described algorithm in C and run it on a Sun Ultra 5/333 with a 256 Mbyte memory. The ISCAS'89 circuits [BBKo89], and the ones known as *Addendum* to the ISCAS'89 benchmark set [Adde93] have been used to evaluate the effectiveness of our approach.

In Tab. 1 we give some experimental results: for each circuit, we first designed the optimal 90/150 linear HCA according to [CaZa95], and then computed the Fault Coverage attained by fault simulating the sequence obtained by making the CA evolve for T=100,000 clock cycles. We then gave these CA rules to the SG algorithm, and obtained new rules that, when implemented into the HCA and fault simulated for the same number of clock cycles, yielded the fault coverage reported in the last column.

Although the BIST circuit runs for 100,000 clock cycles, for efficiency reasons the fitness computation function considers only the first 10,000 vectors generated by each CA. The SG algorithm executes 1,000 iterations, unless premature convergence occurs or a 100% fault coverage is reached earlier, and the best result evaluated is returned. CPU times range from minutes to a few hours per experiment, and are mostly due to fault simulation time.

For sake of comparison, in Table 1 we also reported the results taken from [CCPS97], which concern a higher-overhead BIST architecture (based on a CA composed of two flip-flops per cell) which has been optimized using a standard Genetic Algorithm. Blank cells in this column are due to the fact that the results reported in [CCPS97] did not cover the whole set of circuits considered here.

Table 1: Experimental results

Circuit	Fault Coverage [%]		
	90/150	[CCPS97]	Selfish Gene
s208	94.85	100.0	100.00
s298	23.79	89.26	89.51
s344	61.22	97.49	81.81
s349	61.10	96.98	80.60
s382	16.50	94.27	94.17
s386	42.37	86.00	62.45
s400	15.94	92.37	92.27
s420	87.36	60.03	93.60
s444	13.83	92.43	92.35
s499	71.78		83.93
s510	100.00	100.0	100.00
s526n	11.09	86.12	85.98
s526	11.10	86.03	85.81
s635	0.06		0.06
s641	88.65	88.65	88.70
s713	84.78	84.64	84.78
s820	46.54	52.89	47.92
s832	46.01	52.06	46.76
s838	51.39	34.93	55.74
s938	51.39		57.21
s953	99.46	99.33	99.46
s967	98.89	98.07	98.89
s991	98.47	92.44	98.47
s1196	94.16	89.45	95.30
s1238	91.48	85.46	91.89
s1269	99.79	99.88	99.82
s1423	63.51	86.62	87.56
s1488	56.16	96.59	88.26
s1494	55.95	96.05	92.91
s1512	53.66	70.00	76.26
s3271	99.42	97.78	99.42
s3330	74.96	75.80	78.07
s3384	92.06		93.25
s4863	97.04		97.94
s5378	70.44		75.31

The analysis of these results show that the Selfish Gene algorithm is effectively able to improve, in most of the cases, the industry standard solution, based on the adoption of a 90/150 CA. The generated CA has approximately the same area of the 90/150 one, therefore designers are provided with a much better fault coverage at no additional hardware cost. When compared with the solution proposed in [CCPS97], the current method reaches comparable fault coverage with a much lower area overhead.

5   Conclusions

We described the application of the Selfish Gene algorithm for the solution of a crucial problem in the field of Electronic CAD: the identification of the best structure of a Cellular Automaton in charge of generating the input vectors within a BIST structure.

Experimental results show that the tool is able to identify very good solutions: in fact, with the generated HCA it is possible to reach a Fault Coverage much higher than the one obtained with standard engineering practice. The new approach appears to be very effective even when compared with a previously proposed method based on the usage of a Genetic Algorithm for optimizing a more area expensive CA structure. Testing of embedded circuits is thus made possible with a BIST approach which does not require any intervention from the outside unless test activation and result gathering.

6   References

[ABFr90] M. Abramovici, M.A. Breuer, A.D. Friedman: "Digital Systems Testing and Testable Design," Computer Science Press, 1990

[Adde93] These benchmark circuits can be downloaded from the CAD Benchmarking Laborarory at the address **http://www.cbl.ncsu.edu/www/CBL_Docs/ Bench.html**

[BBKo89] F. Brglez, D. Bryant, K. Kozminski, "Combinational profiles of sequential benchmark circuits," *Proc. Int. Symp. on Circuits And Systems*, 1989, pp. 1929-1934

[BoKa95] S. Boubezari, B. Kaminska, "A Deterministic Built-In Self-Test Generator Based on Cellular Automata Structures," *IEEE Trans. on Comp.*, Vol. 44, No. 6, June 1995, pp. 805-816

[CaZa95] K. Cattell, S. Zhang, "Minimal Cost One-Dimensional Linear Hybrid Cellular Automata of Degree Through 500", JETTA, Journal of Electronic Testing an Test Application, Kluwer, 1995, pp. 255-258

[CCPS97] S. Chiusano, F. Corno, P. Prinetto, M. Sonza Reorda, "Cellular Automata for Sequential Test Pattern Generation", VTS'97: IEEE VLSI Test Symposium, Monterey CA (USA), April 1997, pp. 60-65

[CGPS98] F. Corno, N. Gaudenzi, P. Prinetto, M. Sonza Reorda, "On the Identification of Optimal Cellular Automata for Built-In Self-Test of Sequential Circuits", VTS'98: 16th IEEE VLSI Test Symposium, Monterey, California (USA), April 1998

[CPSo96] F. Corno, P. Prinetto, M. Sonza Reorda, "A Genetic Algorithm for Automatic Generation of Test Logic for Digital Circuits", IEEE International Conference On Tools with Artificial Intelligence, Toulouse (France), November 1996

[CSSq98a] F. Corno, M. Sonza Reorda, G. Squillero, "The Selfish Gene Algorithm: a New Evolutionary Optimization Strategy", SAC'98: 13th Annual ACM Symposium on Applied Computing, Atlanta, Georgia (USA), February 1998, pp. 349-355

[CSSq98b] F. Corno, M. Sonza Reorda, G. Squillero, "A New Evolutionary Algorithm Inspired by the Selfish Gene Theory", ICEC'98: IEEE International Conference on Evolutionary Computation, May, 1998, pp. 575-580

[CSSq99] F. Corno, M. Sonza Reorda, G. Squillero, "Optimizing Deceptive Functions with the SG-Clans Algorithm", CEC'99: 1999 Congress on Evolutionary Computation, Washington DC (USA), July 1999, pp. 2190-2195

[Gold89] D.E. Goldberg, "Genetic Algorithms in Search, Optimization, and Machine Learning," Addison-Wesley, 1989

[HMPM89] P.D. Hortensius, R.D. McLeod, W. Pries, D.M. Miller, H.C. Card, "Cellular Automata-Based Pseudorandom Number Generators for Built-In Self-Test," *IEEE Trans. on Computer-Aided Design*, Vol. 8, No. 8, August 1989, pp. 842-859

[Jett95] JETTA, Journal of Electronic Testing, Theory and Applications, special Issue on Partial Scan Methods, Volume 7, Numbers 1/2, August/October 1995

[KMZw79] B. Konemann, J. Mucha, G. Zwiehoff, "Built-In Logic Block Observation Technique," *Proc. IEEE International Test Conference*, October 1979, pp. 37-41

[KrPi89] A. Krasniewski, S. Pilarski, "Circular Self-Test Path: A low-cost BIST Technique for VLSI circuits," *IEEE Trans. on CAD*, Vol. 8, No. 1, January 1989, pp. 46-55

[NCPa92] T.M. Niermann, W.-T. Cheng, J.H. Patel, "PROOFS: A Fast, Memory-Efficient Sequential Circuit Fault Simulator," *IEEE Trans. on CAD/ICAS*, Vol. 11, No. 2, February 1992, pp. 198-207

[SCDM94] J. van Sas, F. Catthoor, H. De Man, "Cellular Automata Based Deterministic Self-Test Strategies for Programmable Data Paths," *IEEE Trans. on CAD*, Vol. 13, No. 7, July 1994, pp. 940-949

[ToMa87] T. Toffoli, N. Magolus, "Cellular Automata Machines: A New Environment for Modeling," MIT Press, Cambridge (USA), 1987

[Wolf83] S. Wolfram, "Statistical Mechanics of Cellular Automata," *Rev. Mod. Phys.* 55, 1983, pp. 601-644

Dynamic Optimisation of Non-linear Feed-Forward Circuits

Ernesto Damiani[1], Valentino Liberali [2], and Andrea G. B. Tettamanzi[1]

[1] Università degli Studi di Milano, Polo Didattico e di Ricerca di Crema
Via Bramante 65, 26013 Crema, Italy
edamiani@crema.unimi.it, tettaman@dsi.unimi.it
[2] Università degli Studi di Pavia, Dipartimento di Elettronica
Via Ferrata 1, 27100 Pavia, Italy
valent@ele.unipv.it

Abstract. An evolutionary algorithm is used to evolve a digital circuit which computes a simple hash function mapping a 16-bit address space into an 8-bit one. The target technology is FPGA, where the search space of the algorithm is made of the combinational functions computed by cells, of the interconnection topologies and of the interconnections among cells. This circuit is readily applicable to the design of *set-associative* cache memories, with *on-line* tuning of the function during cache operation.

1   Introduction

Evolvable Hardware (EHW) is a relatively recent research field dealing with the synthesis of electronic circuits by means of evolutionary algorithms. The central idea of this approach is representing circuits as individuals in a evolutionary process in which the standard genetic operations can be carried out. At the first stages of EHW research, evolving circuits were often evaluated using software simulation models, while the final outcome had to be implemented in hardware by traditional methods. Today, thanks to the availability of *Field Programmable Gate Arrays (FPGA)*, the evolutionary process may be carried out *on-line*, periodically updating the FPGA structure and content to reflect ongoing evolution.

This work is aimed at on-line evolution of FPGA configurations in order to implement *adaptive hardware components*, i.e. subsystems that automatically modify their structure and operation at the gate and flip-flop level, according to the statistical properties of their input.

The conceptual architecture of an adaptive hardware component is shown in Fig. 1. It includes a *demultiplexer*, sending the input bytestream in parallel to a *working copy* of the adaptive component and to an *evolutionary engine*. The evolutionary engine contains a population of *candidate configurations* for the component and controls their evolution, computing their fitness with respect to the incoming data stream and, possibly, avoiding configurations that are illegal or undesirable for the target implementation [1]. Functionally, the evolutionary engine comprises two subsystems: a *preprocessor* extracting from the input data

J. Miller et al. (Eds.): ICES 2000, LNCS 1801, pp. 41–50, 2000.

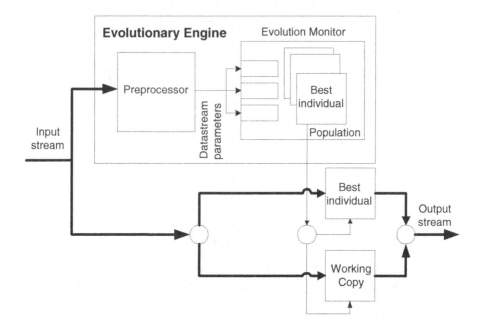

Fig. 1. Conceptual architecture of an adaptive hardware component

stream the parameters relevant to the computation of fitness, and the *evolution monitor* implementing the evolutionary algorithm. The evolutionary engine and the working copy of the circuit operate asynchronously, but the fitness of best individual of the population and the one of the working copy of the configuration are regularly compared. Every time the fitness of the best individual in the population overtakes the fitness of the working copy by a (tunable) threshold, the input data stream is switched from the working copy of the component to the best individual in the population, which becomes the new working copy. This conceptual architecture satisfies the requirement of ensuring uninterrupted operation of the evolvable component while being transparent both to the data and the control flow of the host system. Moreover, it can be instantiated to obtain several classes of components of practical interest. In the sequel, we shall focus on applying this approach to the evolution of a non-linear feed-forward circuit whose input is a stream of *keys*. The circuit computes a non-linear mapping of the keys to a smaller set of *indexes*, adapting the mapping to follow the dynamics of the keys' probability distribution.

2   Problem Formulation

Let $K = \{1, 2, \ldots, M\}$ be a set of keys to be mapped into a set $I = \{1, 2, \ldots, N\}$ of indices in the hash table, with $N \ll M$. Furthermore, assume keys in K have a probability distribution p_k, for all $k \in K$.

Let $\mathcal{H}: K \to I$ be the mapping calculated by a given hash circuit. Then, the probability distribution p_k on K is mapped by $\mathcal{H}$ to the associated probability distribution q_i on the set of indices such that, for all $i \in I$,

$$q_i = \sum_{k:\mathcal{H}(k)=i} p_k. \tag{1}$$

For each index $i \in I$, we can express the number of keys that $\mathcal{H}$ maps into i as $n_i = \sum_{k=1}^{M} [i = \mathcal{H}(k)]$, where notation $[P]$ stands for 1 if predicate P is true and 0 otherwise.

We are interested in finding a hash circuit that maps keys into indices of the hash table in such a way that the most probable (i.e. most frequent) keys have their own entries in the table, without collisions with other keys and that the number of other keys a given key has to share its entry with in the hash table be inversely proportional to its frequency.

The number of other keys a given key $k \in K$ shares an entry with is called the number of collisions c_k, given by $c_k = n_{\mathcal{H}(k)} - 1$. It can be observed that $c_k \geq 0$, because at least key k is mapped into $\mathcal{H}(k)$.

The design goal can be mathematically expressed by saying that we want to minimise the objective

$$z_{\mathcal{H}} = \sum_{k=1}^{M} p_k c_k^a, \tag{2}$$

where a indicates how much penalty we give to collisions as they stack up. For $a = 1$, the cost of each additional collision is constant, while for $a = 2$ it increases quadratically.

Assuming a linear penalty for additional collisions, it is possible to express the objective in Equation 2 in terms of features of the hash table, as follows: $z_{\mathcal{H}} = \sum_{k=1}^{M} p_k c_k = \sum_{i=1}^{N} q_i \sum_{k:\mathcal{H}(k)=i} c_k = \sum_{i=1}^{N} q_i n_i (n_i - 1)$.

In practical situations, we do not know the probability distribution over the set of keys and, therefore, we cannot calculate the q_i's. Besides, even though we could in principle calculate the n_i's, that would take $O(M)$ steps, which is impractical for large values of M. Instead, we rely on a sample of $m < M$ keys with repetitions (which we can think of as extracted from K according with the unknown probability distribution). Based on this sample, we can use the frequency of a key k, $\hat{p}_k$, as an unbiased estimator of its probability, as $E[\hat{p}_k] = p_k$. However, it is more difficult to estimate n_i with the number of keys in the sample mapped to index i, $\hat{n}_i$. If we denote by $\hat{n}_i(m)$ the number of keys out of a sample of size m mapped to index i, it is easy to verify that $E[\hat{n}_i(m)] \leq E[\hat{n}_i(m+1)] \leq n_i$. Nevertheless, one can assume that the missing contributions to $\hat{n}_i$ are from quite unprobable keys, provided a reasonable m is taken, whereby one could conclude that

$$\hat{z}_{\mathcal{H}} = \frac{1}{m} \sum_{i=1}^{N} h_i \hat{n}_i (\hat{n}_i - 1) \tag{3}$$

would provide a reasonable approximation of $z_{\mathcal{H}}$, where h_i is the number of keys in the sample that hit index i, $\sum_{i=1}^{N} h_i = m$. Notice that, by analogy with Equation 1, the frequency $\hat{q}_i$ of index i given a sample of m keys is $\hat{q}_i = \frac{h_i}{m} = \sum_{k:\mathcal{H}(k)=i} \hat{p}_k$.

Now we denote by $z(\mathcal{H}, \mathbf{p})$ the value $z_{\mathcal{H}}$ takes up when keys are distributed according to $\mathbf{p} = (p_1, p_2, \dots, p_M)$. Similarly, we denote by $\hat{z}_m(\mathcal{H}, \hat{\mathbf{p}})$ the approximation to $z_{\mathcal{H}}$ defined in Equation 3 based on a sample of m keys with frequencies $\hat{\mathbf{p}} = (\hat{p}_1, \hat{p}_2, \dots, \hat{p}_M)$. We want to verify that $\hat{z}_m(\mathcal{H}, \hat{\mathbf{p}})$ converges to $z(\mathcal{H}, \mathbf{p})$ as m goes to infinity.

This allows us to write (the easy proof is omitted for the sake of conciseness):

$$\lim_{m \to \infty} \left(\min_{\mathcal{H}} \hat{z}_m(\mathcal{H}, \hat{\mathbf{p}}) \right) = \min_{\mathcal{H}} \left(\lim_{m \to \infty} \hat{z}_m(\mathcal{H}, \hat{\mathbf{p}}) \right) = \min_{\mathcal{H}} z(\mathcal{H}, \mathbf{p}). \tag{4}$$

2.1 Dynamic Key Distribution

Suppose that probability distribution $\mathbf{p}$ changes over time. We are interested in investigating how the objective function changes when $\mathbf{p}$ changes to $\mathbf{p} + \Delta\mathbf{p}$, where, of course, $\Delta\mathbf{p}$ is such that $\sum_{k=1}^{M} \Delta p_k = 0$ because $\mathbf{p} + \Delta\mathbf{p}$ is still a probability distribution. After defining $\Delta\mathbf{q}$ as the change of distribution $\mathbf{q}$ induced by a change $\Delta\mathbf{p}$ in the distribution of keys, and observing from Equation 1 that

$$\Delta q_i = \sum_{k:\mathcal{H}(k)=i} \Delta p_k, \tag{5}$$

we can write

$$z(\mathcal{H}, \mathbf{p} + \Delta\mathbf{p}) = \sum_{i=1}^{N} (q_i + \Delta q_i) n_i (n_i - 1) = z(\mathcal{H}, \mathbf{p}) + \sum_{i=1}^{N} \Delta q_i n_i (n_i - 1). \tag{6}$$

If we define $\Delta z(\mathcal{H}, \mathbf{p}) \equiv \sum_{i=1}^{N} n_i (n_i - 1) \sum_{k:\mathcal{H}(k)=i} \Delta p_k$, Equation 6 can be rewritten as

$$z(\mathcal{H}, \mathbf{p} + \Delta\mathbf{p}) = z(\mathcal{H}, \mathbf{p}) + \Delta z(\mathcal{H}, \mathbf{p}). \tag{7}$$

Given a statistical model of $\Delta\mathbf{p}$, i.e., of the dynamics of key distribution, it can be shown that, on average, this variation is quite small, so that a hash function $\mathcal{H}$ that is optimal for $\mathbf{p}$ has a high probability of being optimal for $\mathbf{p} + \Delta\mathbf{p}$ as well. Therefore, the evolutionary algorithm will be able to track the optimal hash circuit provided the key distribution does not change too fast.

2.2   A Locality-Based Model for Key Distribution Dynamics

In many applications, the distribution over K of Equation 1 will be time-dependent, in a way depending on the stochastic properties of the source. We shall now investigate how the evolutionary process adapting our circuit can be used to track this moving probability distribution in real time. Suppose keys to be addresses generated by a CPU to access memory locations while executing a program. In this case, the source will exhibit a behavior according to the well-known *locality principle* [2]. Specifically, we can expect key probability to be initially distributed according to an exponential law, with a few very probable keys and many little probable keys. Moreover, the number of keys with probability greater than a given threshold increases exponentially as the threshold goes to zero. Accordingly, we generate the initial probability distribution as follows, given a real parameter $0 < u < 1$ governing the *uniformity* of the distribution:

1. $P \leftarrow 1$;
2. $p_k \leftarrow 0$, for all $k \in K$;
3. $k \leftarrow \text{RANDOM}(1, \ldots, M)$;
4. $p_k \leftarrow p_k + (1 - u)P$;
5. $P \leftarrow uP$;
6. if $P \geq \epsilon$, go to 3, otherwise stop.

In Step 6, ϵ is the smallest positive floating-point number that can be represented in double precision.

In addition, we need to provide a law according to which the key distribution changes over time. Again, we model the scope of the locality principle allowing the probability associated to each key to change in two ways:

- by dropping to zero, when the key exits the scope of the locality, i.e. the program execution flow moves away from the address. In this case, another key gets an increase in probability of the same amount
- by averaging its value with the probability of another key. Taking the average models three different situations. First of all, it applies to a key entering the scope of the locality. This happens when one of the involved keys has zero probability, while the other has a non-zero probability value. Secondly, when both involved keys have non-zero probability, taking the average models the fluctuations in probability inside the locality scope. Finally, when both keys have zero probability, it models a null move.

Summarizing the above discussion, we describe an *atomic unit* of probability change as follows:

1. $j, k \leftarrow \text{RANDOM}(1, \ldots, M)$, such that $j \neq k$;
2. flip a fair coin:
3. if *head*, $p_j \leftarrow p_j + p_k$ and $p_k \leftarrow 0$;
4. if *tail*, $p_j, p_k \leftarrow \frac{p_j + p_k}{2}$.

We can adjust the speed of distribution change by specifying how many of these atomic changes happen per unit of time. In fact, the probability change described above leads to a uniform probability distribution in the long run. However, this does not affect our experiments which do no not last so long as to be affected by this long term behavior.

3 Evolutionary Algorithm

A population of hash circuits evolves tracking the changing key distribution thanks
to an evolutionary algorithm whose features are essentially the same described in [3–
5]. That is a simple generational evolutionary algorithm using fitness-proportionate
selection with elitism and linear fitness scaling.

3.1 Encoding and Interconnection Map Evolution

The encoding adopted is an extension of the one described in detail in [5], allowing for
the co-evolution of the interconnection map together with the circuit. For each cell,
the inputs to the cell are encoded in four 16-bit integers, modulo a 64-bit mask of
permissible input cells; the boolean function computed by the cell is encoded as its
16-bit truth table. Altogether, a single cell is encoded by a block of nine 16-bit integers
and 64 such blocks represent a complete hash circuit.

The fitness of a circuit strongly depends on the neighborhood graph [5]. In this
work, circuits are evolved without strict *a-priori* conditions on the neighborhood of
cells. A condition of maximum distance has been imposed, to avoid long connections
which would lead to non-routable circuits. Fig. 2 illustrates the circuit made up of
64 configurable logic blocks (CLB's). Four possible conditions are imposed on the
interconnections: inputs of cells F, G, R, and S may be fed with signals coming from
their neighbouring cells (marked with an "O").

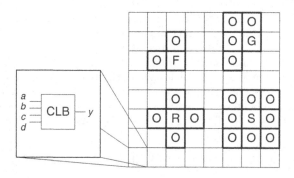

Fig. 2. Four possible boundaries imposed to a cell neighbourhood

The distance d between cells C_{ij} and C_{kl} is defined through a norm. As an example,
for cell S in Fig. 2, the norm is: $d(C_{ij}, C_{kl}) = \max(|i - k|, |j - l|)$ Let us denote the
maximum allowed distance with d_{max}.

Each cell in the circuit has a set of neighbouring cells (*"neighbourhood"*) $\mathcal{M}_{ij}$:
$\mathcal{M}_{ij} = \{C_{kl} \mid C_{kl} \neq C_{ij}; \ d(C_{ij}, C_{kl}) \leq d_{max}\}$ Possible inputs of cell C_{ij} are the
external lines and the outputs of cell belonging to the set $\mathcal{M}_{ij}$.

3.2   Repair Algorithm

The random choice of interconnections may lead to unacceptable solutions. Circuits with loops are not allowed, since they would oscillate or have memory effects. A combinational circuit $\mathcal{C}$ (without loops) is always made up of an ordered set of blocks, i. e. blocks fed with input lines do precede output blocks. The ordering rule is: $C_{ij} < C_{kl}$ (C_{ij} *before* C_{kl}) if the output of cell C_{ij} is fed (either directly or through other cells) to C_{kl}. On the other hand, $C_{ij} > C_{kl}$ (C_{ij} *after* C_{kl}) if the output of cell C_{kl} is fed (either directly or through other cells) to C_{ij}. Two cells C_{ij} and C_{kl} are not related if no signal path includes both of them (partial ordering). All signal paths from input lines to the output must occur through an ordered subset of the circuit cells. Let us denote the set of cells feeding the inputs of cell C_{ij} with $\mathcal{N}_{ij}$. Of course, $\mathcal{N}_{ij} \subseteq \mathcal{M}_{ij}$. Assuming that sets $\mathcal{N}_{ij}$ have been randomly generated by the genetic algorithm, loops may occur and cell ordering is not guranteed. Therefore, it is mandatory either to *repair* circuits contining loops or to discard them.

Repair can be accomplished as follows: after a first attempt at cell ordering, the circuit cells are divided into two subsets: the *ordered* one ($\mathcal{O}$) and the *unordered* one ($\mathcal{U}$), with $\mathcal{O} \cup \mathcal{U} = \mathcal{C}$ and $\mathcal{O} \cap \mathcal{U} = \emptyset$. If $\mathcal{U} = \emptyset$, then all cells have been ordered and there is no loop. Otherwise, a loop exists in the subset $\mathcal{U}$. A cell $C_{ij} \in \mathcal{U}$ is randomly selected, and one of the cells $C_{kl} \in \mathcal{N}_{ij}$ feeding the inputs of C_{ij} is removed from $\mathcal{M}_{ij}$ (and from $\mathcal{N}_{ij}$, of course). Ordering and repair are iterated until all cells of the circuit are in the ordered set $\mathcal{O}$.

3.3   Mutation and Recombination

Whenever an individual is replicated to be inserted in the new generation, each bit of its genotype is flipped with the same probability p_{mut} and independently of the others. Of course, mutation can produce loops in the circuit, for some cell interconnection schemes. If this is the case, loops are subsequently removed by the repair algorithm described above. As in [5], the mutation rate is dynamically tuned while the algorithm is running to maintain a certain level of population diversity. For recombination, each cell is treated as an atomic unit of the genotype, and is inherited from either parent with $\frac{1}{2}$ probability (uniform crossover). This ensures that recombination will preserve the CLBs, which are the basic building blocks of a circuit.

3.4   Fitness and Selection

Minimising the objective in Equation 3 is equivalent to maximising a fitness defined as:

$$f(\mathcal{H}) = \frac{1}{1 + \hat{z}_{\mathcal{H}}} \qquad (8)$$

As it can be readily noticed, such fitness is a random variable, depending on the key sample available when it is calculated. Therefore, the fitness of all individuals is recomputed at each generation, even for those that did not undergo any changes (this is always the case for the best individual). Thus, even though noisy, fitness is always consistent at each generation, being computed on the most recent key sample for all individuals. On average, an unfavourable sample will penalise the fitness of all individuals in the same way, and no bias is introduced into the evolutionary process by fitness noise. Selection probability is proportional to fitness, which is linearly scaled. Elitism is enforced throughout the algorithm.

4   Experiments and Results

In this Section we outline our experimental plan and report our main results. First of all, we observe that, in order to choose the experimentation parameters we relied on a number of previous results. Specifically, in [4] we discussed the feasibility of on-line evolution for this kind of circuits, showing that a small population size can be used for time-bounded computations without compromising the performance of our evolutionary engine. According to these results, we used for the current experimentation a population size of 16 individuals. At each generation, 4096 keys are extracted randomly, according to the current probability distribution, and the performance of each individual is assessed with respect to this sample using the genetic algorithm of Section 3. The size of the sample was chosen to be realistic with respect to implementation in hardware by means of the preprocessor module in Fig. 1. The value of u was set at $\frac{3}{4}$ in order to model a realistically restricted locality scope.

We run our algorithm for different probability distribution dynamics, each characterized by the number of atomic units of key distribution change N_C from one generation to the next. Table 1 reports our results for seven different interconnection schemes: the four schemes illustrated in Fig. 2 (F_1, G_1, R_1, and S_1) and some of their extensions to a longer distance (F_2, G_2, and R_2). Each entry in the table contains the average number of generations at which no collisions occur and the fitness value stably reaches the optimum ($f = 1$), along with the standard deviation. In one case evolution did not converge all the times.

Table 1. Experimental Results

N_C	128		256		512		1024	
	mean	stdev	mean	stdev	mean	stdev	mean	stdev
F_1	9.3	5.7	35.0	34.5	80.0	67.1	38.0	74.8
F_2	8.0	6.2	13.6	14.2	15.0	12.2	26.7	16.5
G_1	3.7	2.7	10.3	5.7	11.4	8.9	22.5	21.2
G_2	7.8	7.0	11.3	9.8	12.3	10.2	15.2	12.5
R_1	15.0	6.8	23.4	32.6	26.0	30.4	35.0	102.3
R_2	15.2	7.3	25.2	28.8	28.9	29.2	n/a	n/a
S_1	5.3	4.9	11.2	8.0	13.2	10.3	24.4	18.2

5   Sample Applications

As an example, we shall describe how a probabilistic hash function module complying with the architecture outlined in Section 1 can be used in order to adaptively associate tags to frame blocks in a *cache memory*.

In microcomputer systems, cache memories are standard, fast access components used to store copies of main memory locations often accessed by the CPU, including their addresses. Each time the CPU generates a memory address, *fully associative* cache memories use the address as an *associative tag*, simultaneously comparing it to all stored addresses (also called *tags*) in the cache memory. This comparison identifies in a single step the "right" cache location, which contains a copy of the content of the corresponding main memory location.

As the size of cache memories increased, it turned out fully-associative cache architectures to be limited by the topological complexity of the combinational network used to perform simultaneous comparisons. In *set-associative* cache memories, a set of *block identifiers* is associated via a combinational network to a smaller set of *tags*. As before, identifiers are (prefixes of) memory addresses provided by the CPU in the framework of a memory Read/Write operation. This time, however, tags do not correspond to single cache locations but to *sets* of locations, i.e. to cache *blocks*. For instance, in a 16-bit linear memory setting, 16-bit identifiers are mapped to a smaller set of 8-bit tags. The cache memory comprises 256 blocks, each corresponding to a tag. Each block entry contains a 8 bit prefix and 16 bits of data, so that blocks hold 512 bytes of data each, giving a total cache capacity of 64 Kbytes. When an identifier (i.e. a 16-bit address) is presented by the CPU at the cache input, a combinational network computes the tag associated to it, identifying the cache block where the value resides. Usually, this computation is performed by taking the most significant byte of the identifier as a block tag. Once the block has been found, a location is accessed inside it, using the least significant part of the identifier to find a match for the location content's prefix. This comparison can be made without excessive performance degradation [6]. In our example, since each cache block contains $l = 256$ entries, accessing the right location inside the block requires an average of $\frac{l}{2} = 128$ comparisons.

Our hash circuit supports a different technique of identifiers-to-tags association, more suitable to *adaptive cache memories*. In an adaptive cache memory, smaller blocks contain locations having high probability of being accessed, according to the probability distribution of identifiers in the input data stream. Since highly probable identifiers correspond to smaller blocks, the number of comparisons required to identify a cache location inside a block is usually shorter than in conventional cache memories. The average number of comparisons to be performed to find a location inside a block is given by

$$\frac{1}{2N} \sum_{i=1}^{N} q_i l_i$$

where N is the number of blocks, q_i is the probability of a key to be stored in the i-th block, and l_i is the length of the i-th block. This formula gives a maximum of $\frac{l}{2}$ when all blocks have an equal length l and $q_i = \frac{1}{N}$ for $i = 1, 2, \ldots, N$. As we have seen, the locality principle tells us that the probability distribution of addresses generated by the CPU changes over time. Therefore, the evolutionary engine must provide the configuration which is the current best for handling the input stream, while evolving other configurations to tolerate change.

6   Conclusion

This paper has presented a novel strategy for dynamic optimization of a hash function circuit. Experimental results show that a small population evolutionary algorithm, which could be in principle implemented in silicon, is able to effectively track a changing key distribution in real time. We intend to compare our experimental approach to available theoretical results about dynamically changing fitness landscapes [7] in order to better justify our choices in parameter settings. Comparison with other techniques

for on-line circuit evolutions is also planned, as soon as suitable experimental data will
be available.

References

1. Levi, D., Guccione, S. A.: GeneticFPGA: Evolving stable circuits on mainstream
   FPGA devices. In Proc. First NASA/DoD Workshop on Evolvable Hardware,
   Pasadena, CA, USA, July 1999. pp. 12–17
2. Patterson, D. A., Hennessy, J. L.: Computer Organization & Design: the Hard-
   ware/Software Interface. Morgan Kaufmann Publishers, San Francisco, CA, USA,
   1994
3. Damiani, E., Liberali, V., Tettamanzi, A. G. B.: Evolutionary design of hashing
   function circuits using an FPGA. In Sipper, M., Mange, D., Pérez-Uribe, A. (Eds.),
   Proc. Second International Conference on Evolvable Systems (ICES '98), Lausanne,
   Switzerland, Sept. 1998. pp. 36–46
4. Damiani, E., Tettamanzi, A. G. B., Liberali, V.: On-line evolution of FPGA-based
   circuits: A case study on hash functions. In Proc. First NASA/DoD Workshop on
   Evolvable Hardware, Pasadena, CA, USA, July 1999. pp. 26–33
5. Damiani, E., Liberali, V., Tettamanzi, A. G. B.: FPGA-based hash circuit synthesis
   with evolutionary algorithms. IEICE Transactions on Fundamentals, Sept. 1999
6. Burkardt, W.: Locality aspects and cache memory utility in microcomputers. Euromicro Journal, vol. 26, 1989
7. Wilke, C., Altmeyer, S., Martinetz, T.: Large-scale evolution and extinction in a
   hierarchically structured environment. In Adami, C., Belew, R. K., Kitano, H.,
   Taylor, C. (Eds.), Proceedings of the 6th International Conference on Artificial Life
   (ALIFE-98), MIT Press, Cambridge, MA, USA, June 27–29 1998. pp. 266–274

From the Sea to the Sidewalk:
The Evolution of Hexapod Walking Gaits
by a Genetic Algorithm

Ernest J P Earon, Tim D Barfoot, and Gabriele M T D'Eleuterio

Institute for Aerospace Studies
University of Toronto
4925 Dufferin Street
Toronto, Ontario, Canada
M3H 5T6
{eea,tdb}@sdr.utias.utoronto.ca, gde@utias.utoronto.ca

Abstract. A simple evolutionary approach to developing walking gaits for a legged robot is presented. Each leg of the robot is given its own controller in the form of a cellular automaton which serves to arbitrate between a number of fixed basis behaviours. Local communication exists between neighbouring legs. Genetic algorithms search for cellular automata whose arbitration results in successful walking gaits. An example simulation of the technique is presented as well as results of application to *Kafka*, a hexapod robot.

1 Introduction

Insects and spiders are examples of relatively simple creatures from nature which are able to successfully operate many legs at once in order to navigate a diversity of terrains. Inspired by these biological marvels, robotics researchers have attempted to mimic insect-like behaviour in legged robots [2][3]. Typically, however, the control algorithms for these types of robots are quite complicated (e.g., dynamic neural networks), requiring fairly heavy on-line computations to be performed in real time. Here a simpler approach is presented which reduces the need for such computations by using a coarsely coded control scheme. This simplicity in the control code (and corresponding low computational requirements on the controller itself) allows for algorithms of this type to be easily implemented using embedded microcontrollers and thus making applications of similar mobile robots more easily realisable. This becomes more appealing if one is interested in making extremely small walkers.

Work has been done to increase the feasibility of using simulation to develop evolutionary control algorithms for mobile robots [11][12][18]. It has been suggested that hardware simulation is often too time consuming, and software simulation can accurately develop control strategies instead. It has been argued that the primary drawback to software simulation is the difficulty in modeling hardware interactions in software. While Jacobi has developed "minimal simulations" [12] in order to develop such controls, it has been found that while

J. Miller et al. (Eds.): ICES 2000, LNCS 1801, pp. 51–60, 2000.

adequate behaviours can be evolved, there are still hardware concerns that are difficult to predict using minimal simulations (e.g. damaging current spikes in the leg drive motors) [10].

This work follows a similar vein to that of [8][9][10]. In [10], Gomi and Ide use a genetic algorithm to develop walking controllers for a mobile robot, an octopod robot. However, rather than reducing the search space through pre-electing a reduced set of behaviours, they very precisely *shaped* [6] the fitness function for reinforcement. One benefit of this is that such fitness function optimization is much more easily scaled up to allow for more complicated behaviour combinations.

What is important to note, though, is that while Gomi and Ide [10] used measurements of the electric current supplied to the leg drive motors as one feature of their fitness function, as opposed to using only the observable success of the controller (i.e., the distance walked as measured with an odometer) as done in the present work, the results have similar attributes. For example, behaviours that could cause a high current flow in the leg motors such as having two or more legs in direct opposition, are also behaviours that would exhibit low fitness values when measuring the forward distance travelled. Thus when the controllers are evolved on hardware such parameters are often taken into account by the robot itself, and modeling is not necessarily required.

As with dynamic neural network approaches, each leg of a robot is given its own controller rather than using a central pattern generator. Communication between controllers is local (legs "talk" to their neighbours) resulting in a global behaviour (walking gait) for the robot. There is some neurobiological evidence that this notion of local control holds for some biological insects [4]. The controllers used here are cellular automata [16] or simple lookup tables. Each leg is given its own cellular automaton (CA) which can be in one of a finite number of states. Based on its state and those of its neighbours, a new state is chosen according to the lookup table. Each state corresponds to a fixed *basis behaviour* but can roughly be thought of as a specific leg position. Thus, new leg positions are determined by the old leg positions. Typically, all legs are updated synchronously such that the robot's behaviour is represented by a set of locally coupled difference equations which are much quicker to compute in real time than differential equations.

Under this framework, design of a control algorithm is reduced to coming up with the cellular automata which produce successful walking gaits. Our approach has been to use genetic algorithms to this end [15].

2   Cellular Automata

According to neurobiological evidence described by Cruse [4], the behaviour of legs in stick insects is locally coupled as in Figure 1. This pattern of ipsilateral and contralateral connections will be adopted for the purposes of discussion although any pattern could be used in general (however, only some of them would be capable of producing viable walking gaits).

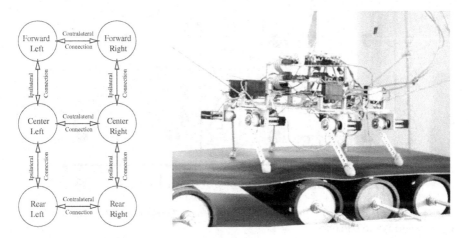

Fig. 1. (left) Behavioural coupling between legs in stick insects [4]. (right) *Kafka*, a hexapod robot, and treadmill setup designed to evolve walking gaits

We assume that the output of each leg controller may be discrete. This may be done by way of a set of *basis behaviours* [1][13]. Rather than specify the actuator positions (or velocities) for all times, we assume that we may select a simple behaviour from a finite predefined palette. This may be considered a postprocessing step which takes a discretized output and converts it to the actuator control. This postprocessing step will not be allowed to change once set. The actual construction of the postprocessing requires careful consideration but is also somewhat arbitrary. Here the basis behaviours will be modules which move the leg from its current zone (in output space) to one of a finite number of other zones. Figure 2 shows two possible discretizations of a 2-degree-of-freedom output space (corresponding to a simple leg) into 4 or 3 zones. It is important to distinguish between a basis behaviour and a leg's current zone. If legs always arrive where they are asked to go, there is little difference. However, in a dynamic robot moving on rough terrain, the requested leg action may not always be successful. The key difference is that the *output state* of a leg will correspond to the current basis behaviour being activated, rather than the leg's current zone in output space. The *input state* of a leg could be either the basis behaviour or leg zone, depending on whether one wanted to try and account for unsuccessful requests. By using basis behaviours, the leg controllers may be entirely discrete. Once all the postprocessing has been set up, the challenge remains to find an appropriate arbitration scheme which takes in a discrete input state, x, (current leg positions or basis behaviours of self and neighbours) and outputs the appropriate discrete output, a, (one of M basis behaviours) for each leg. There are several candidates for this role but the one affording the most general decision surfaces between input and output is a straightforward lookup table similar to *cellular automata* (CA) In fact, this work follows on that of the Evolving Cellular Automata (EvCA) group at the Santa Fe Institute, New Mexico. In various papers

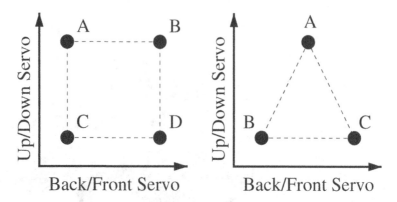

Fig. 2. Example discretizations of output space for 2 degree of freedom legs into (left) 4 zones and (right) 3 zones

(e.g., [5]) this group showed that genetic algorithms were able to evolve cellular automata which performed prescribed tasks requiring global coordination. This is essentially what we wish to achieve but we have the added difficulty of dealing with the physical environment of our robots. This type of lookup table control in autonomous robots is often called *reactive*. For every possible input sequence the CA scheme stores a discrete output value. In other words, for every possible input sequence there is an output corresponding to one of the basis behaviours. At each time-step, the leg controller looks up the action which corresponds to its current input sequence and carries it out. The size of the lookup table for a leg which communicates with $K - 1$ other legs will then be M^K. The approach is therefore usually rendered feasible for only modest numbers for K and M. The number of all possible lookup tables is $M^{(M^K)}$. Again, modest numbers of basis behaviours keep the size of the search space reasonable. For example, with a hexapod robot with coupling as in Figure 1 and output discretization as in Figure 2 (left) the forward and rear legs will require lookup tables of size 4^3 and the central legs 4^4. If we assume left-right pairs of legs have identical controllers the combined size of the lookup tables for forward, center, and rear legs will be $4^3 + 4^3 + 4^4 = 384$ and the number of possible table combinations for the entire robot will be 4^{384}. From this point on, the term *CA lookup table* will refer to the combined set of tables for all legs in the robot (concatenation of individual leg lookup tables).

3 Genetic Algorithms

The crucial step in this approach is the discovery of particular CA lookup tables which cause the connected legs to produce successful walking gaits. We must discover the local rules which produce the desired global behaviour, if they indeed exist, and then find out why those rules stand out. The obvious first method to attempt is to design the local rules by hand. John Horton Conway picked his

Game of Life rules seemingly out of a hat with much success. How hard can it be? Well, it turns out to be very difficult to do this for all but the most trivial examples. We are typically faced with a combinatorial explosion. In the spirit of the EvCA group, an evolutionary global optimization technique will be employed, namely a genetic algorithm (GA), to search for good cellular automata.

GAs are based on biological evolution. For a good review see [7]. A random initial population of P CA lookup tables is evolved over G generations. Each CA lookup table, ϕ, has a *chromosome* which consists of a sequence of all the discrete values taken from the table. At each generation, a fitness is assigned to each CA lookup table (based on how well the robot performs on some walking task when equipped with that CA lookup table). A CA lookup table's fitness determines its representation in the next generation. Genetic crossovers and mutations introduce new CA lookup tables into the population. The best $K \leq P$ CA lookup tables are copied exactly from one generation to the next. The remaining $(P - K)$ CA lookup tables are made up by single site crossovers where both parents are taken from the best K individuals. Furthermore, they are subjected to random site mutations with probability, p_m, per site. The variable used to determine mutation is selected from a uniform random distribution.

4   Example Controller

In this section, the results of constructing an example controller in simulation will be presented. The purpose of the simulation is not to model an insect robot, but rather to demonstrate that a network of cellular automata controllers could produce patterns which resemble known walking gaits. Furthermore, we would like to show that a genetic algorithm is indeed capable of finding particular CA lookup tables which perform well on a user defined task.

One well established gait for hexapod robots is the "tripod" gait. The legs are divided into two sets

$$\{\text{Forward Left}, \text{Center Right}, \text{Rear Left}\}$$
$$\{\text{Forward Right}, \text{Center Left}, \text{Rear Right}\}$$

While one set is on the ground, the other is lifted, swung forward, and then lowered.

For the discretization of figure 2 (left), there are $I = 4^6 = 4096$ possible initial states for a hexapod robot. The fitness of a particular CA lookup table will be defined as

$$f_{total} = \frac{\sum_{i=1}^{I} f_i}{I} \tag{1}$$

$$f_i = \begin{cases} 1 & \text{if tripod gait emerges within } T \text{ time-steps} \\ & \text{starting from initial condition, } i \\ 0 & \text{otherwise} \end{cases} \tag{2}$$

This fitness function was used to evolve a particular CA lookup table named ϕ_{tripod}, which has a fitness of 0.984. That is, from 98.4% of all possible initial conditions, a tripod gait is reached. In our experiments, we have used a GA population size of $P = 50$, number of generations $G = 150$, keepsize $K = 15$, and mutation probability $p_m = 0.005$. The number of initial conditions per fitness evaluation was $I = 4096$ (all possible initial conditions) and the number of time-steps before testing for a tripod pattern was $T = 50$. Figure 3 depicts

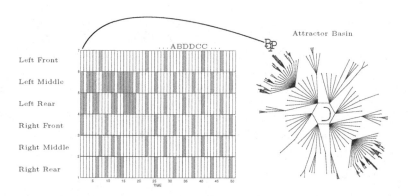

Fig. 3. (left) Gait diagram (time history) of ϕ_{tripod} on a particular initial condition. Each 6-state column represents an entire leg configuration of the robot; the left-most column is the initial condition. (right) Attractor basin portrait of ϕ_{tripod}. Each node represents an entire 6-state leg configuration of the robot. Lines represent transitions from one leg configuration to another. The inner hexagon represents the $ABDDCC\ldots$ tripod gait.

two aspects of ϕ_{tripod}. The left side shows a typical *gait diagram* or *time history* of ϕ_{tripod} on a particular initial condition. Each column shows the states of the 6 legs at a given time-step (different shades of grey represent different leg states). The left-most column is the initial condition. The right side is an attractor basin portrait of ϕ_{tripod} as described by [17]. Each node in this plot represents an entire 6-state leg configuration of the robot (i.e., one column of the left plot). The inner hexagon represents the tripod gait which is unidirectional (indicated by a clockwise arrow). The purpose of the right plot is to draw a picture of ϕ_{tripod} as a whole and to make a connection with the concept of stability. Beginning from any node on the attractor basin portrait, and following the transitions inward (one per time-step), one will always wind up on the inner hexagon (tripod gait).

The conclusion we may draw from this simple exercise is that this type of controller certainly is able to produce patterns resembling known walking gaits and that genetic algorithms are capable of finding them.

5   Hardware and Results

This section briefly describes *Kafka* [14], a 12-degree-of-freedom, hexapod robot. Figure 1 shows Kafka in action on a treadmill. Kafka's legs are powered by twelve JR Servo NES 4721 hobby servomotors controlled by a 386 66MHz PC. The control signals to the servos are absolute positions to which servos then move as fast as possible. In order to control the speed of the leg motions, then, the path of the leg is broken down into many segments and the legs are sent through each of those. The fewer the intervening points, the faster the legs moved. The robot has been mounted on an unmotorized treadmill in order to automatically measure controller performance (for walking in a straight line only). As Kafka walks, the belt on the treadmill causes the rollers to rotate. An odometer reading from the rear roller is fed to Kafka's computer such that distance versus time-step plots may be used to determine the performance of the controller. The odometer measures net distance travelled. For this experiment, the states for Kafka's legs were constrained to move only in a clockwise, forward motion which is based on the three state model as shown in figure 2 (right) (i.e., rather than having each leg move from one state to an arbitrary next state, it is constrained to either increment through the sequence of states in the clockwise motion, or to not change states). Each entry in the lookup table determines whether to increment the position of the leg or keep it stationary (binary output). This reduces the number of possible lookup tables from 3^{135} (2.58×10^{64}) for the full three-state model to 2^{135} (or 4.36×10^{40}). The GA parameters were

$$P = 50 \qquad K = 15 \qquad p_m = 0.05 \qquad (3)$$

Each individual in the population was given the same initial condition, or stance, at the start of the evaluation ($\{C, B, C, B, C, B\}$).

Due to the necessary time duration required to evolve the controllers on Kafka, the number of generations that the GA was run was 23 (the evolution of the controllers on Kafka took approximately eight hours to complete). The convergence history for one evolution (population best fitness and population average fitness vs. generation) is shown in Figure 4. Figure 5 shows the gait diagrams for four of the best individuals from various generations (ϕ_{one}, ϕ_{two}, ϕ_{three}, ϕ_{four}, from generations 3, 5, 13, and 22 respectively, with ϕ_{four} being the best gait generated during the GA run). All four of the gaits performed fairly well, walking in a consistently forward direction during their evaluations. ϕ_{one} corresponds to the first sharp increase in fitness from previous generations. The improvement between ϕ_{one} and ϕ_{two} is due to the shorter period for the step cycle and the gait pattern itself is simpler. This trend continues in ϕ_{three} and ϕ_{four} with a much reduced step cycle period and a more consistent and simplified gait pattern. The fitness improvements corresponding to these gaits is indeed quite high compared to the earlier two individuals.

Figure 4 shows the fitness generated by each individual throughout the entire GA run. The rather high fitness value reported in generation 3 (Figure 4) is a result of an anomalous sensor output. This was confirmed when, in the next

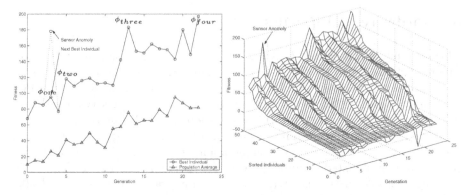

Fig. 4. Convergence History of a GA Run. (left) Best and average fitness plots over the evolution. (right) Fitness of entire population over the evolution (individuals ordered by fitness). Note there was one data point discounted as an odometer sensor anomaly. In the best individual convergence history it was replaced by the next best individual (as indicated on left graph).

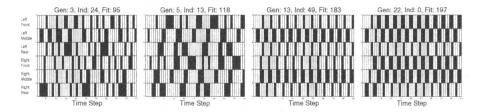

Fig. 5. Gait diagrams (time histories) for the four solutions, $\phi_{one}, \phi_{two}, \phi_{three}, \phi_{four}$, respectively.

generation, that individual achieved an extremely low fitness (-17). The rest of the plot shows quite clearly the trend of increasing fitness for the kept segment of the population as well as the variance in fitness reported for different individuals. Due to the difficulty in running large populations on a hardware apparatus such as Kafka, the population size is somewhat low. However, as can be seen in Figure 4 by this fitness variance, the genetic diversity is maintained.

As mentioned earlier, the number of generations over which the GA was allowed to evolve was rather limited. This can been seen in Figure 4 by the fact that there is still a rather high deviation for the best individuals during the latter generations. The average fitness would not necessarily converge to the values reported for the best individuals, even if those gaits had saturated the population. This is due to the fact that the mutation rate of $p_m = 0.05$ is sufficiently high to keep many of the child individuals in each generation from being simply copies of the parents.

It should be pointed out that fitness does not monotonically increase with generation. There is a fluctuation in the fitness of the individuals after convergence. This is due to the hardware aspect of the experiment. Even the best K

individuals were re-evaluated each generation. As the experiment progressed, the rubber foot pads used by Kafka simply wore out. This loss of footing corresponded to a variable loss in traction and fitness for each gait.

The use of a state table with binary output greatly reduced the search space for this experiment. The next stage for this type of work involves using a full state lookup table (3 or more output possibilities). The expected result of this is the same as for the reduced version, however, the time to develop successful gaits would be increased, and the convergence history would likely be much more gradual (likely making it impractical). For example, if there were 9 output zones for each leg, the chromosome would have length 8019 and the search space would be of size 9^{8019}. Thus, our approach may not scale very well to increases in model complexity (due to the long training times required).

6   Conclusion

Many researchers believe that highly parallel and decentralized methods are the key to endowing artificial systems with intelligence. Decentralized controllers for insect robots offer a great deal of redundancy in that if one controller fails, the robot may still limp along under the power of the remaining functional legs [3]. The cellular automata controller approach outlined here was able to successfully control Kafka, a hexapod robot, and should extend to robots with more degrees of freedom (keeping in mind scaling issues). One advantage of using such a coarse controller is that it requires very few real-time computations to be made (compared to dynamic neural network approaches) as each leg is simply looking up its behaviour in a table. The approach easily lends itself to automatic generation of controllers through genetic algorithms (or any global optimization technique) as was shown for the simple examples presented here. Although it would be costly to evolve walking gaits on real robots (inverse design by optimization is always computational expensive), this may be done off-line (ahead of time). This work was motivated by evidence that biological insects generate walking patterns by means of decentralized control [4]. We hope that studying such types of control for artificial walkers may also in turn tell us something about the natural systems by which they were inspired.

Acknowledgments

Kafka, the hexapod robot featured in this work, was constructed by David McMillen as part of a Master's thesis at the University of Toronto Institute for Aerospace Studies.

The authors would like to thank Thierry Cherpillod for much help in setting up the experiment as well as the Natural Sciences and Engineering Research Council of Canada and the Canadian Space Agency for supporting this work.

References

1. T D Barfoot and G M T D'Eleuterio. An evolutionary approach to multiagent heap formation. Congress on Evolutionary Computation, July 6-9 1999.
2. Randall D. Beer, Hillel J. Chiel, and Leon S. Sterling. A biological perspective on autonomous agent design. *Robotics and Autonomous Systems*, 6:169–186, 1990.
3. Hillel J. Chiel, Randall D. Beer, Roger D. Quinn, and Kenneth S. Espenschied. Robustness of a distributed neural network controller for locomotion in a hexapod robot. *IEEE Transactions on Robotics and Autonomation*, 8(3):292–303, june 1992.
4. Holk Cruse. Coordination of leg movement in walking animals. In J. A. Meyer and S. Wilson, editors, *Simulation of Adaptive Behaviour: from Animals to Animats*. MIT Press, 1990.
5. Rajarshi Das, James P. Crutchfield, Melanie Mitchell, and James E. Hanson. Evolving globally synchronized cellular automata. In L.J. Eshelman, editor, *Proceedings of the Sixth International Conference on Genetic Algorithms*, pages 336–343, San Fransisco, CA, April 1995.
6. Dorigo and Colombetti. *Robot Shaping*. MIT Press, A Bradford Book, 1998.
7. David E. Goldberg. *Genetic Algorithms in Search, Optimization, and Machine Learning*. Addison-Wesley Pub. Co., Reading, Mass., 1989.
8. Takashi Gomi. Practical applications of behaviour-based robotics: The first five years. In *IECON98, Proceedings of the 24th Annual Conference of the IEEE*, pages 159–164, vol 4. Industrial Electronics Society, 1998.
9. Takashi Gomi and Ann Griffith. Evolutionary robotics - an overview. In *Proceedings of IEEE International Conference on Evolutionary Computation*, pages 40–49, 1996.
10. Takashi Gomi and Koichi Ide. Evolution of gaits of a legged robot. In *The 1998 IEEE International Conference on Fuzzy Systems*, pages 159–164, vol 1. IEEE World Congress on Computational Intelligence, 1998.
11. Freédéric Gruau. Automatic definition of modular neural networks. *Adaptive Behaviour*, 3(2):151–184, 1995.
12. Nick Jakobi. Runnig across the reality gap: Octopod locomotion evolved in a minimal simulation. In P Husbands and J-A Meyer, editors, *Proceedings of Evorob98*. Evorob98, Springer-Verlag, 1998.
13. Maja J. Matarić. Behaviour-based control: Examples from navigation, learning, and group behaviour. *Journal of Experimental and Theoretical Artificial Intelligence*, 9(2):232–336, 1997. Special Issue on Software Architectures for Physical Agents, Editors: H. Hexmoor, I. Horswill, D. Kortenkamp.
14. David Ross McMillen. Kafka: A hexapod robot. Master's thesis, University of Toronto Institute for Aerospace Studies, 1995.
15. Richard S. Sutton and Andrew G. Barto. *Reinforcement Learning: An Introduction*. A Bradford Book, MIT Press, 1998.
16. Jon von Neumann. *Theory of Self-Reproducing Automata*. University of Illinois Press, Urbana and London, 1966.
17. Andrew Wuensche. The ghost in the machine: Basins of attraction of random boolean networks. In Chris G. Langton, editor, *Artificial Life III: SFI Studies in the Sciences of Complexity, vol. XVII*. Addison-Wesley, 1994.
18. David Zeltzer and Michael McKenna. Simulation of autonomous legged locomotion. In Chris G. Langton, editor, *Artificial Life III: SFI Studies in the Sciences of Complexity, vol. XVII*. Addison-Wesley, 1994.

Experiments in Evolvable Filter Design using Pulse Based Programmable Analogue VLSI Models

Alister Hamilton[1], Peter Thomson[2] and Morgan Tamplin[1]

[1] Department of Electronics and Electrical Engineering,
University of Edinburgh, King's Buildings,
Mayfield Road, Edinburgh EH9 3JL, Scotland.
Alister.Hamilton@ee.ed.ac.uk
[2] School of Computing,
Napier University, 219 Colinton Road,
Edinburgh EH14 1DJ, Scotland.
p.thomson@dcs.napier.ac.uk

Abstract. The direct evolution of hardware, thus far, has concentrated largely upon the design of small, feed-forward digital designs. In this paper, a procedure is described, and results of experiments discussed, for evolving analogue filters using integrator cell models based upon working programmable analogue VLSI cells developed by one of the authors. This, of necessity, involves using feedback interconnects and settings in the range 0.0 - 1.0 which represents, for each individual cell, the gain (ratio of output power to input power) of the integrator. Experiments are conducted in the direct evolution of low-pass filter systems, and the implications of the results discussed. The evolutionary model used is similar in structure to that described in the companion paper [7].

Introduction

The field of evolvable hardware is now becoming established with the attempts of [1, 3, 2, 5, 8, 6] to directly evolve electronic systems. Our contribution to this discipline is to demonstrate the use of novel techniques in programmable analogue and mixed signal VLSI as hardware platforms for evolution.

To this end, this paper describes experiments in the design of sampled data analogue filters using a software model of the pulse based programmable analogue VLSI cells developed by one of the authors and described in [9]. The basic programmable analogue cell is a differential integrator with an electronically programmable gain. When evolving these designs it is possible to use a genetic algorithm (GA) chromosome to describe the interconnect (in both the feedforward and the feedback sense) between cells and their corresponding gains. A chromosome therefore describes a complete filter design. Our first attempt at evolving sampled data filter structures passes a series of signals through the filter at incremental frequencies. The fitness then is the measure of how close the

J. Miller et al. (Eds.): ICES 2000, LNCS 1801, pp. 61–71, 2000.
© Springer-Verlag Berlin Heidelberg 2000

outputs from the filter at these different frequencies match the desired or target response of the idealised filter.

In the following section we give a brief overview of the programmable analogue VLSI cell and describe how filters may be constructed from these functional blocks. We then go on to discuss results of experiment to directly evolve first, second, third and fourth order low pass filters.

The Programmable VLSI Environment

Research at the University of Edinburgh has led to novel techniques for implementing programmable analogue and mixed signal VLSI. The relatively simple and novel leap taken at Edinburgh of representing an analogue sampled data signal by the width of a digital pulse rather than a more conventional current or voltage mode representation has opened up a range of new possibilities [9].

The basic programmable analogue cell is a differential integrator whose input and output analogue signals are represented by width modulated digital signals and whose gain (K) is electronically programmable (Figure 1). The internal operation of the cell is entirely analogue. Integrators were chosen as the programmable cell as they may be used for a large number of analogue signal processing functions including filtering. The power of the technique lies in the

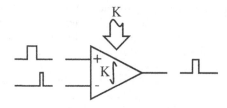

Fig. 1. Programmable analogue cell with electronically programmable gain (K) factor. Analogue input and output signals are represented by the width of a digital pulse.

fact that programmable cells may be easily interconnected using digital signal routing techniques, that mixed signal processing is easily performed as outlined in [9] and that continuous time as well as sampled data signal processing is possible.

There are a wide variety of design techniques available for the synthesis of analogue filters. Many date back to the early days of electronics before the introduction of active devices like the transistor and the operational amplifier. These techniques make use of passive components; inductors (L), capacitors (C) and resistors (R) to make filters. Due to the mathematical operations performed by these components on time varying signals, networks of L, C and R components may be directly translated into networks of integrators with different gains. One example of such a circuit is shown in Figure 2. In general, the order of the filter is determined by the number of integrators used. The higher the order of the filter, the better it is at discriminating between signals that the filter should pass and signals that the filter should attenuate.

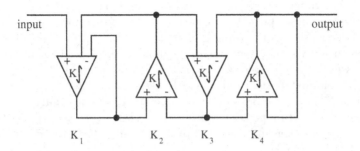

Fig. 2. 4th order filter implemented using differential integrators.

As we have stated, filter design is nothing new and many books and tables of filter designs and co-efficients exist. Our purpose here is to develop experimental techniques in order to synthesise filters using a model of the programmable analogue cell. We see this as a first step in designing circuits with more complex functionality that make use of both the analogue and digital components that our programmable VLSI techniques make available. While these initial experiments should provide analogue circuit structures that are familiar, later experiments are likely to produce mixed signal circuits that have not been seen before.

The Model

There are many different levels at which we can model the functionality of the programmable analogue cell. At one extreme we could model the cell at the detailed transistor level using a simulation tool such as SPICE, but this would take too much simulation time. At the other extreme we could model the cell as a purely mathematical function, but this would not map well into our future plans for mixed signal circuits. We have chosen an intermediate model where pulses are represented by a real number and added to, or subtracted from a real number representing the integrated value depending on the sign of the signal and the interconnectivity. The output pulse width is defined as the integrated value multiplied by a real valued scaling factor.

The Experiment

The GA used in these experiments to perform cellular evolution uses uniform crossover at a rate of 100% and mutation at a rate of 10%. Selection is done on a tournament basis (tournament size 2) with a tournament probability of 70%. Throughout a population size of 10 is used initially over a run of 50 generations. This increases to a maximum of 500 generations when evolving higher order filters. The first part of the chromosome consists of a list of integers which represent the cell inter-connect net-list and the output node for the filter. The remaining part of the chromosome consists of a list of real numbers which represent the individual integrator cell gains in the range 0 to 1.0. This data set completely describes a phenotype which is a pulse based filter.

A full description of the cellular array structure used in terms of geometry and interconnectivity is given in a companion paper [7]. The original cellular structure was entirely feed-forward, but this has been modified in these experiments to accommodate the feedback paths essential to filter design.

A set of 11 input vectors were generated to represent a range of sinusoidal input frequencies with a peak magnitude of 1. The frequencies represented were relative to one another and consisted of a range from 0.1fc to 1.5fc where fc is closely related to the desired cut-off frequency of the low pass filter.

The fitness is calculated by measuring the mean square error between the target response (defined in a file) and the actual output response from each of the individual filters created by the GA chromosomes. Clearly the idea is to minimise this error function.

It is anticipated that later experiments will present a step input to each evolved filter and the filter step response will be evaluated using a fast Fourier transform (FFT). The fitness will then be determined by a comparison of the actual and desired frequency spectra.

We attempted to evolve first, second, third and fourth order filter structures by incrementally increasing the number of cells available in the cellular array. For example, a single cell system would represent a first order filter, but increasing this to two cells creates the potential to evolve a second order system. Increasing the order of the filter will result in a better match to the target response indicated by a smaller fitness value.

It is further possible to use cell lock-down [7] to preserve both the connectivity and gain of individual cells whose behaviour is deemed to have influence upon the fitness. This approach may be used to seed a further run of the GA in order to derive fitter results over a shorter evolutionary cycle. In other words, we are setting a good starting point for subsequent evolution to be more successful than a random population selection.

Experimental Results

Experiments with a Single Cell

The first experiment used a single cell. The aspiration behind this experiment was that a negative feedback connection would naturally evolve between the output of this cell and the inverting (minus) input. This took relatively few generations (maximum 50) to evolve and all experiments were able to establish a negative feedback connection. A typical circuit is shown in Figure 3 and the response of the fittest run is shown in Figure 4. The fitness value was 1.089. Cell lock-down was used to preserve the connectivity and gain of the cells and

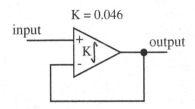

Fig. 3. Evolved first order filter with evolved gain (K) factor.

the experiment was re-run. The fitness improvement obtained in this experiment was marginal with a final value of 1.087. This result is also shown in Figure 4.

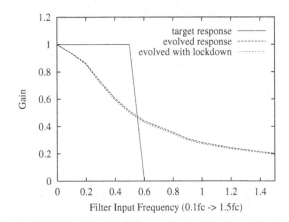

Fig. 4. Target response for filter. Response from the evolved first order filter. Response after lock-down has been applied for a further run of the GA.

Experiments with Two Cells

This experiment used two pulse based integrator cells. Here the hope was that two integrator cells would be interconnected in such a way as to further reduce the fitness value obtained in the first experiment, thereby producing a magnitude response that more closely matched the target. 10 runs of the GA with a maximum of 50 generations for each run were used to evolve second order structures. The best of these, with a fitness value of 0.764, is shown in Figure 5 with corresponding gain (K) values. The magnitude response of the evolved filter is shown in Figure 6. Cell lock-down was used to seed the connectivity and gain co-efficients for a second set of GA runs. An improved fitness value of 0.541 was achieved with identical connectivity and the co-efficients shown in brackets in Figure 5.

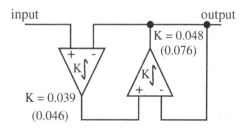

Fig. 5. Fittest evolved second order filter. Evolved filter gain co-efficients are shown. The gain co-efficients in brackets are those obtained by running the GA after lock-down.

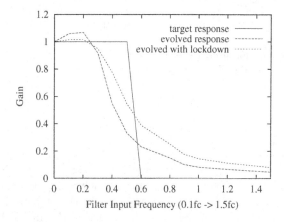

Fig. 6. Target response for filter. Response from the evolved second order filter. Response after lock-down has been applied for a further run of the GA.

Experiments with Three Cells

This experiment used three pulse based integrator cells in an attempt to improve fitness further by evolving a third order filter structure. The number of generations was increased from 50 to 500 for this experiment. A number of interconnection topologies were evolved, that giving the best fitness of 0.396 is shown in Figure 7. The magnitude response of the evolved filter is shown in Figure 8.

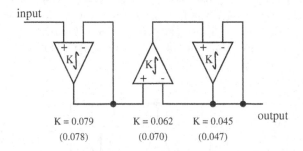

Fig. 7. Fittest evolved third order filter. Evolved filter gain co-efficients are shown. The gain co-efficients in brackets are those obtained by running the GA after lock-down.

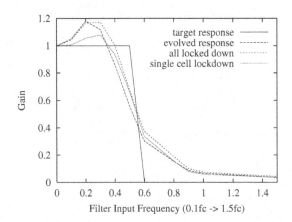

Fig. 8. Target response for filter. Response from the evolved third order filter. Response after lock-down has been applied for a further run of the GA. Response with lock-down applied to a single integrator.

Cell lock-down was applied and a fitness of 0.362 was achieved with the circuit of Figure 7 and the cell co-efficients shown in brackets in this Figure. A different lock-down strategy was tried with the a single integrator locked down. The integrator pins were assigned different nodes, with the input signal going to the non-inverting (plus) input. A fitness of 0.278 was achieved with the connectivity of Figure 7 and gain factors of 0.049, 0.082 and 0.045. The response of this filter is also shown in Figure 8.

Experiments with Four Cells

This experiment used four pulse based integrator cells. Here again 500 generations were used. Interestingly, the best filter structure evolved in this experiment was the classical third order filter structure shown in Figure 9. A fitness value of 0.225 was achieved in one run after 209 generations. After lock-down was applied

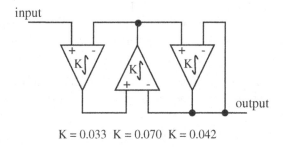

K = 0.033 K = 0.070 K = 0.042

Fig. 9. Fittest evolved filter using populations of four cells and 500 generations. Evolved filter gain co-efficients are shown.

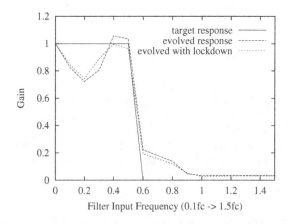

Fig. 10. Target response for filter. Response from the evolved third order filter using populations of 4 cells. Response after lock-down has been applied for a further run of the GA. Response with lock-down applied to a single integrator.

and the GA re-run, a fitness of 0.160 was achieved. The magnitude response for both these results is shown in Figure 10.

Finally, 100 runs of 500 generations were run with lock-down applied to the last result. A fourth order filter structure was evolved with a fitness value of 0.130. The circuit diagram is shown in Figure 11

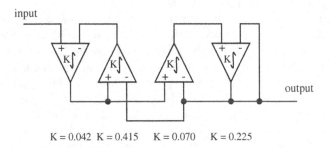

K = 0.042 K = 0.415   K = 0.070   K = 0.225

Fig. 11. Fittest evolved filter using populations of four cells and 500 generations over 100 runs. Evolved filter gain co-efficients are shown.

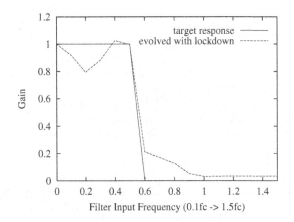

Fig. 12. Target response for filter. Response from the evolved fourth order filter using populations of 4 cells.

Conclusions

We have described a procedure for the direct evolution of analogue filters using a pulse-based VLSI cell model, the actual hardware for which has already been developed at the University of Edinburgh. We have achieved this by developing a software model of the integrator cells which are incorporated on the actual chip, then using accumulated real numbers to simulate the actual circuit behaviour. Using a series of data samples to represent the range of input frequencies, we have shown that it is possible using this method to evolve classical first, second and third order filter structures. This probably represents the limit of the current technique due to the limitation in defining the target response, although an unusual fourth order structure which has been evolved.

The next step in this work is to investigate the possibility of applying an input step function to the evolved filters, and to use a FFT to ascertain the output spectrum. This would address the limitations in the current evaluation method in that all instances in the output would be continuously considered as opposed to the current discrete approach which has only examined eleven individual frequency points. Further, the viability of using these evolved models on the actual hardware needs to be examined, plus the possibility of evolving even higher order filters within some desired operational envelope. However, perhaps more excitingly, there is the possibility of combining these analogue cells into mixed-signal systems where digital cells will also be present. This should give rise to the development of systems that were previously unknown. The next step here is to refine the existing evolutionary model in order that this particular feature may be accommodated.

References

1. Higuchi T., Iwata M., Kajitani I., Iba H., Hirao Y., Furuya T., and Manderick B., Evolvable Hardware and Its Applications to Pattern Recognition and Fault-Tolerant Systems, in Lecture Notes in Computer Science - Towards Evolvable Hardware, Vol. 1062, Springer-Verlag, 1996., pp. 118-135.
2. Koza J. R., Andre D., Bennett III F. H., and Keane M. A., Design of a High-Gain Operational Amplifier and Other Circuits by Means of Genetic Programming, in Evolutionary Programming VI, Lecture Notes in Computer Science, Vol. 1213, Springer-Verlag 1997, pp. 125-135.
3. Thompson A., An evolved circuit, intrinsic in silicon, entwined with physics, in Higuchi T., Iwata M., and Liu W., (Editors), Proceedings of The First International Conference on Evolvable Systems: From Biology to Hardware (ICES96), Lecture Notes in Computer Science, Vol. 1259, Springer-Verlag, Heidelberg, 1997., pp. 390-405.
4. Thompson A., On the Automatic Design of Robust Electronics Through Artificial Evolution., in Sipper, M., Mange, D. and Perez-Uribe, A. (Editors), Proceedings of The Second International Conference on Evolvable Systems: From Biology to Hardware (ICES98), Lecture Notes in Computer Science, Vol. 1478, Springer-Verlag, 1998, pp. 13-24.
5. Miller J. F., Thomson P., and Fogarty T. C., Designing Electronic Circuits Using Evolutionary Algorithms. Arithmetic Circuits: A Case Study., in Genetic Algorithms and Evolution Strategies in Engineering and Computer Science: D. Quagliarella, J. Periaux, C. Poloni and G. Winter (eds), Wiley, 1997.
6. Lohn J. D., and Colombano S. P., Automated Analog Circuit Synthesis Using a Linear Representation., in Sipper, M., Mange, D. and Perez-Uribe, A. (Editors),

Proceedings of The Second International Conference on Evolvable Systems: From Biology to Hardware (ICES98), Lecture Notes in Computer Science, Vol. 1478, Springer-Verlag, 1998, pp. 125-133.

7. Thomson P., Circuit Evolution and Visualisation. submitted to ICES2000 The Third International Conference on Evolvable Systems: From Biology to Hardware., 2000.

8. Goeke M., Sipper M., Mange D., Stauffer A., Sanchez E., and Tomassini M., Online Autonomous Evolware, in Higuchi T., Iwata M., and Liu W., (Editors), Proceedings of The First International Conference on Evolvable Systems: From Biology to Hardware (ICES96), Lecture Notes in Computer Science, Vol. 1259, Springer-Verlag, 1997, pp. 96-106.

9. Hamilton A., Papathanasiou K. A., Tamplin M. R., Brandtner T., Palmo : Field Programmable Analogue and Mixed-Signal VLSI for Evolvable Hardware., in Sipper, M., Mange, D. and Perez-Uribe, A. (Editors), Proceedings of The Second International Conference on Evolvable Systems: From Biology to Hardware (ICES98), Lecture Notes in Computer Science, Vol. 1478, Springer-Verlag, 1998, pp. 335-344.

The Intrinsic Evolution of Virtex Devices Through Internet Reconfigurable Logic

Gordon Hollingworth*, Steve Smith, and Andy Tyrrell

University of York, Heslington, York, England
gsh100,sls,amt@ohm.york.ac.uk
http://www.amp.york.ac.uk/external/media/

Abstract. This paper describes the implementation of an evolvable hardware system on the Virtex Family of devices. The evaluation of the circuits is done intrinsically by downloading a portion of the bitstream that describes the changes from the baseline circuit. This reconfiguration system is achieved through the use of the Xilinx JBits API to identify and extract the changing bitstream. This partial reconfiguration process is very much faster than the previous methods of complete reconfiguration, leading to the complete evolution of a simple circuit in less than half an hour.

1 Introduction

Genetic Algorithms were first explored by John Holland [3]; he showed how it is possible to evolve a set of binary strings which described a system to which a measure of fitness can be applied. The analogy to evolution in nature is that the binary strings are analogous to the DNA sequence (the genotype) carried by all living things, and that the system under test is the phenotype (the body) which is built through embryological development. The body (system) is then subject to the environment where its fitness for reproduction is assessed; fitter individuals have a higher rate of reproduction. This means that genes within the DNA that code for specific 'good' traits (traits which describe better reproduction abilities) will have a higher probability of existing in future populations.

In artificial evolution a binary string describes the system. This system is described directly or through some form of mapping. The system is then assessed within the environment to determine its fitness relative to other individuals. This measure is then used to weight the probability of reproduction so that after many generations, good genes (components of the system) survive and bad genes die away.

A simple example might be to find the maximum of some function f(x); the genotype-phenotype mapping is simple since the value of x to insert into the function is the binary value of the genotype. The fitness is simply the value of the function, and after a number of generations the average fitness increases.

---

* Supported by Engineering and Physical Sciences Research Council

J. Miller et al. (Eds.): ICES 2000, LNCS 1801, pp. 72–79, 2000.

Holland showed that this works mathematically when the rate of reproduction is higher for fitter individuals[3].

A further extension to the domain of evolutionary techniques transpired after the arrival of the Field Programmable Gate Array (FPGA)[9] devices and the Programmable Logic Device (PLD), since these can be programmed using a binary string or by coding a binary logic tree. The electronic circuits can then be evaluated electronically, by comparing their output with that required, (intrinsic Evolvable Hardware (EHW)), or by simulation (extrinsic EHW), to achieve a measure of the fitness[20].

This has promoted much research into EHW [15, 14, 11, 8, 4, 5] demonstrating not only the successful evolution of electronic circuits, but also some highly desirable features, such as fault tolerance [16, 6]. One of the most interesting areas is that of intrinsic evolvable hardware where the circuit is evaluated within the FPGA rather than through simulation[17, 2, 7, 14]. This enables the fitness to be assessed more quickly than in simulation and also allows the use of parallel fitness evaluation.

Much of the work on intrinsic evolution has been applied on custom built hardware [14, 8, 18], using non-mainstream devices like the Xilinx 6200 chip. This paper describes work being carried out that uses a new family of FPGA's that can provide intrinsic evolution on general purpose machines for general purpose applications.

The potential of bio-inspired electronics when applied as adaptive hardware is clearly demonstrated. It can change its behaviour and improve its performance while executing in a real physical environment (as opposed to simulation). Such on-line adaptation is more difficult to achieve, but theoretically provides many advantages over extrinsic systems. At present, work has mostly been concerned with off-line adaptation. That is, the hardware is not used in an execution mode while evolving. Problems involved with on-line adaptation include the time to adapt and the accuracy of adaptation. However, if these problems can be overcome bio-inspired electronics has many potential applications.

2   Virtex

The majority of intrinsic evolvable hardware systems have used the Xilinx XC6200 series chips. This chip has a number of advantages for the evolvable system design.

Microprocessor Interface The interface is very much faster than other FPGA configuration methods, also it is possible to access the configuration memory in a random manner.

Partial Reconfiguration The random access interface means that small changes can be made to the FPGA's configuration.

Safe Configuration The configuration system prevents configurations that can damage the chip, therefore allowing random configurations to be written to the configuration memory.

Unfortunately in May 1998 Xilinx announced its intention to stop producing the devices and consequently the chips have become hard to obtain. Subsequently Xilinx has introduced the new Virtex family, which has many of the properties of the XC6200 series:

SelectMAP interface A Microprocessor compatible interface which is capable of running at up to 66MHZ.

Partial reconfiguration The Virtex device can be partially reconfigured, but there is a minimum size for the programming.

Unsafe Configuration The Virtex device must NOT be programmed with a random bitstream, this can easily destroy the device

Since the device can be damaged with a random configuration bitstream (and Xilinx have not released the bitstream format) it has been generally assumed that evolvable hardware was not possible using the Virtex device. However, with the introduction of the Xilinx JBits API alternative strategies are possible.

The Virtex family [10] uses a standard FPGA architecture as in figure 1. The logic is divided into an NxM structure of Configurable Logic Blocks(CLB), each block contains a routing matrix and two slices. Each slice contains two Lookup Tables (LUTs) and two registers (in addition to considerable amounts of logic not used in these examples). The inputs to the slice are controlled through the routing matrix which can connect the CLB to the neighboring CLBs through single lines (of which there are 24 in each direction) and hex lines which terminate six CLBs away. Each of the inputs to the CLB can be configured using the JBits API classes, as well as all the routing. This means that it is possible to route two outputs together creating a bitstream which will damage the device. If it was also possible to program the device from outside the computer, damage could be terminal, something that the Java security was developed to prevent.

3   JBits

JBits [19] is a set of Java classes which provide an Application Program Interface (API) into the Xilinx Virtex FPGA family bitstream. This interface operates on either bitstreams generated by Xilinx design tools, or on bitstreams read back from actual hardware. It provides the capability of designing and dynamically modifying circuits in Xilinx Virtex series FPGA devices. The programming model used by JBits is a two dimensional array of Configurable Logic Blocks (CLBs). Each CLB is referenced by a row and column, and all configurable resources in the selected CLB may be set or probed.

Since the JBits interface can modify the bitstream it is still possible to create a 'bad' configuration bitstream; for this reason it is important to make sure that the circuit which is downloaded to the chip is always a valid design. This is done by only ever modifying a valid design to implement the evolvable hardware. This is a method similar to that demonstrated in Delon Levi's work on GeneticFPGA in [12] where the XC4000 series devices were programmed to implement evolvable

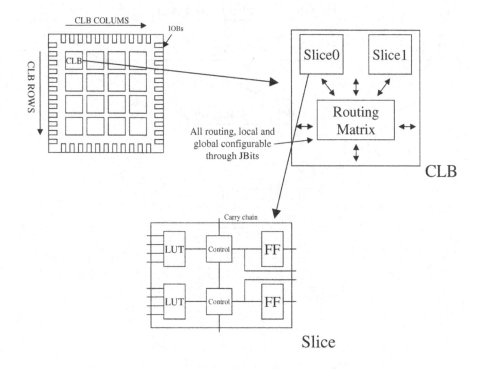

Fig. 1. Virtex Architecture

hardware. The major obstacle in that implementation was the speed with which the devices could be programmed.

The hardware interface is the XHWIF (Xilinx Hardware InterFace). This is a Java interface which uses native methods to implement the platform dependent parts of the interface. It is also possible to run an XHWIF server, which receives configuration instructions from a remote computer system and configures the local hardware system. This also enables the evolvable hardware to be evaluated in parallel on multiple boards. This interface is also part of Xilinx's new Internet Reconfigurable Logic (IRL) methodology, where the computer houses a number of Xilinx parts which can easily be reconfigured through the Java interface. The intention is to create hardware co-processors which can speed up computer systems. Our aim is to use this interface to create evolvable co-processors; evolving solutions to improve the speed of the programs while they are running.

3.1   Partial Reconfiguration through JBits

It is possible to partially reconfigure the hardware system using the JBits interface. This is shown diagrammatically in figure 2. This is accomplished by first identifying the particular region of the bitstream which requires reconfigu-

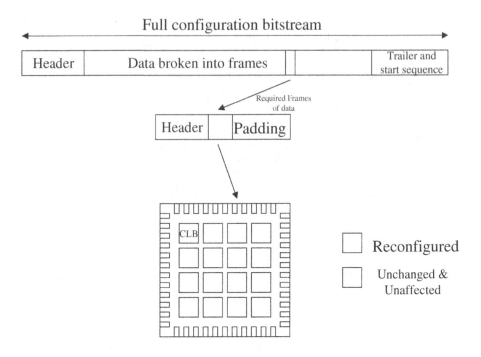

Fig. 2. Dynamic reconfiguration in action

ration and wrapping a header around this bitstream data to identify the address at which to start reconfiguration. During the reconfiguration process the unchanged logic runs as normal with no interruption to the circuit. This means that we can implement the interface to the logic within the circuit, knowing that changes can be made inside the chip without affecting the interface.

4   JAVA GA

The circuit upon which the evolvable hardware is overlaid, is shown in figure 3. The bidirectional bus is used to read and write data to test the logic circuit. The four by two array of LUTs are where the evolvable logic is placed. Note that the routing is all feed-forward, meaning that no loops can be created. Once the bitstream for this circuit is loaded into the JBits class, modification can begin, setting and resetting the bits in the LUTs from the genotype. The genetic algorithm used was written by Steve Hartley from Drexel University [1]. This is a simple genetic algorithm which calls an evaluation function to evaluate the genotype, a number of boolean bits. These bits are split up to create the 16 bits required per LUT, which are then programmed into the bitstream using JBits. The required part of the bitstream (the frames that are actually changed) are then identified and a new bitstream is created to partially reconfigure that part

Population Size	100
Maximum generations	10000
Crossover Probability	0.7
Number of Crossover points	1
Mutation Probability	0.001
Mutation type	Point Mutation

Table 1. Genetic Algorithm Properties

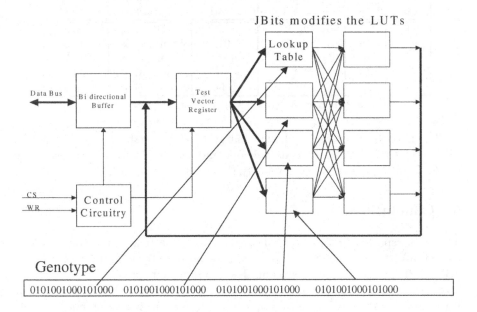

Fig. 3. Virtex Baseline circuit

of the Virtex configuration. The new configuration is then sent to the chip and tested by inputting test vectors and reading the corresponding output vectors.

Figure 3 also shows how the genotype is translated into the LUT values. The GA properties are shown in table 1

The implementation was a simple two bit adder, where the inputs to the circuit are the two 2bit numbers and the expected output is the sum of the two input values. To fully test the circuit 16 input vectors are applied, and the results are read for each input vector and compared to the required output. The fitness is incremented if the value is correct, leading to a maximum fitness of 16.

5   Results

Table 2 shows statistics for five runs of the 2bit adder implementation. The number of evaluations per second show that the whole design can be reconfigured at approximately 40 times per second. This is using a fairly slow interface, the ISA bus at 8 MHz, to program the chip through the SelectMAP interface. In theory the Virtex device can be programmed at up to 66MHz, bringing the average evaluation time down to 3 minutes and 50 seconds. Evolution times of this magnitude permit real systems to evolve in real time.

Run Number	Final Fitness	Final Average Fitness	Number of Generations	Time required to find solution	Evaluations per second
1	16	12.98	657	27 mins 2 secs	40.5
2	16	12.31	1011	39 mins 33 secs	42.6
3	16	13.58	1129	43 mins 2 secs	43.7
4	16	13.45	497	20 mins 31 secs	40.4
5	16	12.65	744	29 mins 11 secs	42.5

Table 2. Five complete runs of the 2bit adder experiment

The main bottlenecks for the current system are the Java Virtual Machine (JVM) which is not as fast as native code and the ISA bus which can only run at 8MHZ, although we hope to remedy this is the near future.

6   Conclusions

The work reported in this paper describes the first attempts to evolve circuits using the Virtex device, which has been made possible through the release of the JBits API for the Virtex Bitstream. The main advantage of the Virtex device over the previous 6200 devices is its price and accessibility. With the previous version of the JBits software (for the XC4000 series devices) it is now possible to evolve circuits for as little as a hundred dollars, making evolvable hardware widely accessible.

The new Virtex-E device has a maximum reconfiguration speed of 66MHz, and with the possibility of Java co-processors and the Internet Reconfigurable Logic interface, the electronic circuits can be evolved on different parallel computer systems just as easily as on the host computer, but at many times the speed.

The next stage in the design is to change the baseline circuit which is used for evaluations, as the circuit does not currently have any form of feedback or register (memory) capability. One variation of baseline circuit that is currently being used is a simple feed-forward network similar to that described in [13]. The major advantage of this type of circuit is that it can be evaluated and compared to that designed by hand.

References

1. Steve Hartley. Java ga. http://www.mcs.drexel.edu/ shartley/.
2. I. Harvey, P. Husbands, and D. Cliff. Seeing the light: Artificial evolution, real vision. In *From Animals to Animats 3, Proceedings of the third international conference on simulation of adaptive behaviour*, 1994.

3. J.H. Holland. *Adaptation in Natural and Artificial Systems.* University of Michigan Press, 1975.

4. G.S. Hollingworth, S.L. Smith, and A.M. Tyrrell. Design of highly parallel edge detection nodes using evolutionary techniques. In *Proceedings of 7th Euromicro Workshop on Parallel and Distributed Processing.* IEEE Press, 1999.

5. G.S. Hollingworth, S.L. Smith, and A.M. Tyrrell. Simulation of evolvable hardware to solve low level image processing tasks. In Poli et al., editor, *Evolutionary Image Analysis, Signal Processing and Telecommunications*, volume 1596 of *LNCS*, pages 46–58. Springer, 1999.

6. G.S. Hollingworth, S.L. Smith, and A.M. Tyrrell. To evolve in a changing environment. In *IEE Colloquium on Reconfigurable Systems.* IEE Informatics, March 1999.

7. H. Iba, M. Iwata, and T. Higuchi. Machine learning approach to gate-level evolvable hardware. In *International Conference on Evolvable Systems: From Biology to Hardware.* Springer, 1996.

8. H. Iba, M. Iwata, and T. Higuchi. *Gate-level Evolvable Hardware: Empirical study and application*, pages 259–279. Springer-Verlag, 1997.

9. Xilinx inc. Xc6200 field programmable gate array data book, 1995. http://www.xilinx.com/partinfo/6200.pdf.

10. Xilinx inc. Virtex field programmable gate arrays databook, 1999. http://www.xilinx.com/partinfo/ds003.pdf.

11. M. Iwata, I. Kajitani, H. Yamada, H. Iba, and T. Higuchi. A pattern recognition system using evolvable hardware. In *International Conference on Evolutionary Computation: The 4th Conference on Parallel Problem Solving from Nature*, pages 761–770. Springer, 1996.

12. D. Levi and S. Guccione. Geneticfpga: Evolving stable circuits on mainstream fpga devices. In *The First NASA/DoD Workshop on Evolvable Hardware.* IEEE Computer Society, 1999.

13. J.F. Miller. Evolution of digital filters using a gate array model. In Poli et al., editor, *Evolutionary Image Analysis, Signal Processing and Telecommunications*, volume 1596 of *LNCS*, pages 17–30. Springer, 1999.

14. M. Murakawa, S. Yoshizawa, I. Kajitani, T. Furuya, M. Iwata, and T. Higuchi. Hardware evolution at functional level. In *International conference on Evolutionary Computation: The 4th Conference on Parallel Problem Solving from Nature*, pages 62–71, 1996.

15. A. Thompson. *Evolving Electronic Robot Controllers that exploit hardware resources.*, pages 640–656. Springer-Verlag, 1995.

16. A. Thompson. Evolutionary techniques for fault tolerance. *UKACC International Conference on Control*, pages 693–698, 1996.

17. A. Thompson. An evolved circuit, intrinsic in silicon, entwined with physics. In *Procedures of the 1st international conference on Evolvable systems (ICES96).* Springer, 1996.

18. A. Thompson. Temperature in natural and artificial systems. In P. Husbands and I. Harvey, editors, *Proc. 4th Int. Conf. on Artificial Life (ECAL97)*, pages 388–397. MIT Press, 1997.

19. Xilinx. Jbits documentation, 1999. Published in JBits 2.0.1 documentation.

20. X. Yao and T. Higuchi. Promises and challenges of evolvable hardware. In *International Conference on Evolvable Systems: From Biology to Hardware.* Springer, 1996.

Evolution of Controllers from a High-Level Simulator to a High DOF Robot

G. S. Hornby[1], S. Takamura[2], O. Hanagata[3], M. Fujita[3], and J. Pollack[1]

[1] Computer Science Dept., Brandeis University, Waltham, MA
{hornby, pollack}@cs.brandeis.edu
http://www.demo.cs.brandeis.edu
[2] ER Business Incubation Dept., Sony Corporation, 6-7-35, Kitashinagawa,
Shinagawa-ku
Tokyo, 141-0001 JAPAN
takam@pdp.crl.sony.co.jp
[3] Group 1, D-21 Lab, Sony Corporation, 6-7-35, Kitashinagawa, Shinagawa-ku
Tokyo, 141-0001 JAPAN
{hana, fujita}@pdp.crl.sony.co.jp

Abstract. Building a simulator for a robot with many degrees of freedom and various sensors, such as Sony's AIBO[4], is a daunting task. Our implementation does not simulate raw sensor values or actuator commands, rather we model an intermediate software layer which passes processed sensor data to the controller and receives high-level control commands. This allows us to construct a simulator that runs at over 11000 times faster than real time. Using our simulator we evolve a ball-chasing behavior that successfully transfers to an actual AIBO.

1 Introduction

One area of evolutionary robotics is evolving controllers for robots. The two approaches are to evolve with actual robots or to evolve controllers in simulation for use with an actual robot. Real robots are always a better model than a simulator, yet they are limited to going at real time. Simulation also has the advantage of being more convenient to work with because of the ease with which performance can be monitored and the robot and world can be reset to perform multiple trials. Simulation's shortcoming is that because it is an imperfect model, controllers created for simulated robots often perform differently on actual robots. Nevertheless there have been several instances of controllers evolved in simulation that have transferred to real robots: locomotion for a hexapod [Gallagher & Beer, 1992] and [Gallagher et al., 1996]; obstacle avoidance with a Khepera [Michel, 1995]; and others.

One approach to developing a simulator is to create a simulator based on data taken from a real robot, [Miglino et al., 1995] and [Lund & Miglino, 1996]. Actual robot sensor readings are used to create lookup tables. In their experiments they evolved a neural control system for a Khepera to move around a

---

[4] AIBO is a registered trademark of Sony Corporation

J. Miller et al. (Eds.): ICES 2000, LNCS 1801, pp. 80–89, 2000.
© Springer-Verlag Berlin Heidelberg 2000

simple environment while avoiding obstacles. Limitations of this methodology are that it lacks a way of developing an accurate model of the physical dynamics of the world, it is only useful for creating a partial model of the sensors, nor would it scale well to more complex sensors, such as a digital camera.

Another approach to constructing a simulator is that of minimal simulations [Jakobi, 1998]. This methodology identifies features of the environment that are easy to simulate and are necessary and sufficient for the robot to perform the desired behavior. These features are reliably simulated for each trial. Other aspects of the simulation are varied for each trial so as to be unreliable. Controllers evolve using only those features of the simulator which were reliably modeled and thus transfer successfully to an actual robot. Successful transference from simulators constructed using this method has been achieved for various tasks with various robots, such as T-maze navigation with a Khepera, locomotion and obstacle avoidance with a hexapod.

Projects using the simulators listed above connected the robot controllers directly to the robot's sensors and actuators. For AIBO, controllers sit on top of sensor processing and locomotion modules. Consequently our simulator does not model raw sensor values or motor commands, instead it models the processed sensory data and effects of high level movement commands. By modeling processed sensor input and high level motor commands our simulator was much easier to create and faster to execute, achieving a speedup of over 11000.

The main motivation in constructing this simulator is for evolving controllers for actual AIBOs. Controllers evolved in simulation should perform similarly on an actual AIBO. To evolve controllers that will transfer we include an error value for sensor values and locomotion similar to that used in minimal simulations [Jakobi, 1998]. Error values are determined randomly at the start of each trial and fixed for the entire trial – as opposed to adding random noise throughout a trial. In this way evolved controllers will be tolerant to sensors and actuators performing in a small range about the simulator setting, yet the controllers will not depend on noise. Using our simulator we evolve a ball-chasing behavior for AIBO which successfully transfers to a real AIBO.

The rest of the paper is organized as follows. In section 2 we describe AIBO and the simulator. Section 3 contains a description of the control architecture being evolved. In section 4 we describe our experiments and results. Finally, in section 5 we summarize our work.

2   Simulator

In figure 1 we show one way of visualizing a robot system. Layer a consists of the high level controller, which is what is evolved in these experiments. Layer b is an intermediate processing layer between the controller and layer c, the robot's sensors and actuators. Layer d is the environment in which the robot exists.

In evolutionary robotics, controllers are typically connected directly to a robot's sensors and actuators, with no layer b. Simulators developed for such systems must then model sensor values and actuator movements. Modeling these

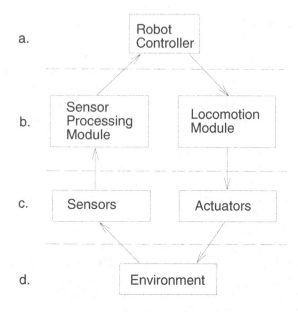

a.

b.

c.

d.

Fig. 1. Decomposition of a robot system.

parts of the system can require much development time and may require a high degree of accuracy, resulting in a slow simulation. Fortunately, AIBO comes with a layer *b* that processes sensor readings and converts high level locomotion commands to control the motors. Thus our simulator does not need to deal with various complex sensors and 14 motors, but need only model sensors in the way that the sensor processing module returns data – providing size and direction to each object instead of determining the color of each pixel of the digital camera – and it models the results of locomotion commands, such as *step forward* instead of modeling the movement of four legs, each with 3 DOF. On an SGI O2 with an R10000 processor our simulator performs approximately 27800 time-steps a second which, at 0.42s per simulated time-step, is a speedup of 11700 – adding processing for the neural controller reduced this to a speedup of 4500.

2.1   AIBO and its Environment

Our simulator models AIBO, a quadruped robot with 18 degrees-of-freedom and various sensors. We use 14 of AIBO's 18 degrees-of-freedom which are controlled through high-level commands, such as *look forward, look down* and which type of locomotion step to take. The sensing capabilities of AIBO that we utilize are the color detection tables of its digital camera, the infrared distance sensor, and joint values of the neck through various modules. Finally, the environment consists of a Robocup-like field and ball. A description of AIBO's prototype can be found in [Fujita & Kitano, 1998].

Walking and running is controlled through the locomotion module. This module takes high-level commands requesting different steps, *forward, front left, front right, turn left, turn right* and *backwards,* and moves AIBO appropriately. Gaits are developed by hand or are evolved [Hornby et al., 1999].

The two sensors we use are a distance sensor and a digital camera. The raw signal from the distance value is converted to a distance value in meters. The images from the digital camera are processed into 8 color detection tables, CDTs. For each table the total number of pixels containing the target color are given, as is the center of the pixel mass. One CDT is assigned to each object (own goal, opponent's goal, ball, teammate and opponent) in AIBO's environment – in these experiments only the CDT data for the ball is used.

The soccer field consists of a flat playing area bordered by walls. At each end a goal-area is marked by a section of colored cloth.

2.2   Simulator

Our simulator consists of a physical-dynamics component and a sensing component. The physical dynamics module handles movement of objects in the world as well as detecting and handling collisions between objects. The sensing component models the data sent by the sensor processing software to the robot's controller.

Fig. 2. Screen shot of the simulator, 1-on-1 with two AIBOs, two goals and a ball. Not shown are the walls lining the sides of the field.

Each AIBO robot is simulated as consisting of a body and a head. The different steps of the locomotion module are modeled as moving AIBO at a given linear and angular velocity. If AIBO's current linear or angular velocity are not the same as the specified value for the type of step being taken, acceleration is used to adjust these velocities according to classical dynamics. Similarly, AIBO's position is updated using classical dynamics.

At the end of each time step objects are checked for collisions. A collision occurs when two objects are found to intersect each other. When a ball collides with an object its velocity is instantaneously changed to its reflection. In addition, the following adjustments to ball velocity are used to model the effects of a ball colliding with a robot: if the ball collided with the front of an AIBO, the ball's velocity is increased by 10-80% of AIBO's linear velocity; if the ball collides with any of the other 3 sides of AIBO, the angle of its linear velocity is changed by up to 20% and its speed is reduced by up to 40%. When an AIBO collides with a wall it is moved perpendicularly away from the wall until it is just touching the wall. When two AIBOs collide with each other, they are moved away from each other along the line through their centers, figure 3, a distance such that they are no longer penetrating. In both types of collisions the linear velocity of the colliding robot(s) is reduced to 0.

a.   b.

Fig. 3. Collision handling: when two robots intersect each other, the collision is handled by moving them apart along the line through their centers.

Modeling of sensing uses error to create uncertainty as with the methodology of minimal simulations. Before each trial the size of the error is decided at random for each sensor. The range of error is ±10% of the sensor value, a range in which the settings for an actual AIBO should lie. Once selected, this amount of error is used throughout the entire trial. A different error is selected for subsequent trials. This forces evolved controllers to perform robustly within the specified tolerance range, hence they should transfer to an actual AIBO. By not implementing the error as a random noise to be added throughout a trial, controllers do not evolve to become dependent on noise.

The most complex sensor to model is the digital camera, which is situated at the front of AIBO's head. A color detection table (CDT) is associated with each object – ball, own goal, opponent goal, teammate, opponent – in the world. For each color detection table, the CDT module returns the size, number of pixels of that color, as well as an and tilt angles to the center of the object visible.

Simulation of the CDT consists of calculating, for each object, the pan and tilt angles to the object's center and its distance from the digital camera. First the distance from the digital camera to the object is calculated. The object's leftmost and rightmost points are then calculated using the distance value along with the object's relative location and orientation from the camera. These values are clipped to be within the 52° field of view of the digital camera. Next, objects are compared for occlusion, with nearer objects occluding farther objects. The exception is that the ball does not obscure anything because of its small size. Once the left and right angles of all objects are known, the center is the average of these two angles. Object size, a measure of how large an object appears on the

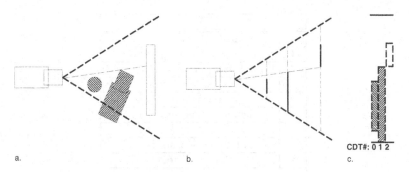

Fig. 4. Simulation of vision, showing the sequence of going from simulated world, to determining how large objects will appear on the robot's CDTs.

robot's digital camera, is calculated from the leftmost and rightmost boundaries on the CDT (see figure 4).

Once vision is completed the infrared distance sensor is handled. The distances from the front of head to the nearest wall, the floor, and any object within ($\pm 5°$) from a direct line of sight are compared. The smallest distance is the one returned as the distance value.

Modeling of both distance and vision sensing takes into account height of the digital camera and pan and tilt angles of the head.

3   Neural Control Architecture

The control architecture we use is a recurrent neural networks similar to that of [Hornby & Mirtich, 1999]. Neural networks consist of an input layer, a number of hidden units and an output layer. The neural controller receives input from AIBO's sensory module and sends commands to the locomotion module.

A neural controller consists of 12 input units, 20 hidden units and 7 output units. The data structure for a neural controller consists of: w, a matrix of real-valued weights; f, a vector of processing functions, $\{x, sin(x), asin(x), \frac{1}{1+e^{-x}}\}$; θ, a real-valued vector of input biases; and τ, a real-valued vector of time constants. The value of the ith row and jth column of the weight matrix is the value of the weight from the ith processing unit to the j processing unit. The ith value of the vectors is the attribute for the ith processing unit. A processing unit functions as follows:

$$a_{i,t} = \tau_i a_{i,t-1} + (1 - \tau)[f(\sum_j w_{ji} a_j + \theta_i)] \tag{1}$$

As units are updated sequentially, the activation value a_j will be $a_{j,t}$ for values of j less than i and $a_{j,t-1}$ otherwise.

The twelve units of the input layer are listed in table 1. The first two units are the angles of the pan and tilt joints of AIBO's neck. The third unit is the distance from AIBO's distance sensor. The next five inputs correspond to stepping: forward; turn left; turn right; forward left; and forward right. The unit for

Unit	Value
0	tilt angle of neck (radians).
1	pan angle of neck (radians).
2	distance from infrared sensor (m).
3	last step was forward.
4	last step was turn left.
5	last step was turn right.
6	last step was front left.
7	last step was front right.
8	ball is visible.
9	size of ball in CDT.
10	tilt angle to ball.
11	pan angle to ball.

Table 1. Inputs to Neural Controller

whichever step was taken by AIBO in the previous time step has a value of 1, the other units have a value of 0. The remaining units are for visual sensing of the ball with AIBO's color detection table. There are units for: indicating whether or not the object is visible; object's size; pan angle to the object; and tilt angle to the object. If an object is not visible, then the reported size, pan angle and tilt angle are those for the last time the object was visible. Other inputs, such as vision information for the goals and other robots, are available but are not used in these experiments.

Unit	Value
0	stop
1	step forward
2	turn left
3	turn right
4	step front left
5	step front right
6	head position

Table 2. Outputs from Neural Controller

The 7 output units are listed in table 2. Outputs 0 through 5 are used to determine the locomotion behavior of AIBO. Whichever of these units has the highest activation value is the type of step that AIBO takes. Unit 6 determines the head position of AIBO. If the activation of this unit is greater than, or equal to, 0 then AIBO's head is moved to the *forward* position (-0.4 radians); otherwise the head is moved to the *down* position (-0.9 radians).

4   Experimental Results

To test our simulator we use it to evolve a neural controller for ball-chasing. Networks are recurrent neural networks with 12 inputs, 20 hidden units and 7 output units. and are evolved the same way as in [Hornby & Mirtich, 1999]. In these experiments we use a population size of 60 individuals, with 60% recombination, 40% mutation and an elitism of 2 individuals. Fitness is the average score over 12 trials – for each trial the score is a function of the distance which the ball is moved over a period of 21 (simulated) minutes.

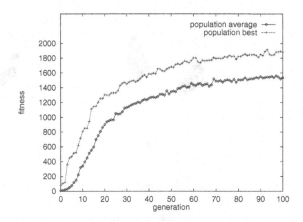

Fig. 5. Graph of average fitness for the entire population and fitness of the best individual, averaged over 6 evolutionary runs. Fitness is distance (cm) the ball is moved over a 21minute (3000 time step) trial.

Figure 5 contains a graph plotting the average fitness of 6 evolutionary runs. Individuals start with very poor ball chasing skills. After a few generations individuals are able to push the ball a few hundred centimeters over the course of 21 minutes. By the end of 100 generations individuals can push the ball more than 1500cm.

Evolved individuals chased the ball in a variety of ways. Initially they would turn until they could see the ball and move towards it. Some would push the ball a little, then re-center on it. Others would give a hard push and, once they lost sight of the ball, would then loop around until they spotted the ball again. All controllers performed the behavior well. Figure 6 contains screen shots of a chase sequence.

Individuals that were transferred to an actual AIBO were successful in chasing a ball. As in the simulated world they generally avoided the wall and were able to relocate the ball after it is moved to a different location in the world. Figure 7 shows pictures of an actual AIBO performing ball chasing with an evolved neural network controller.

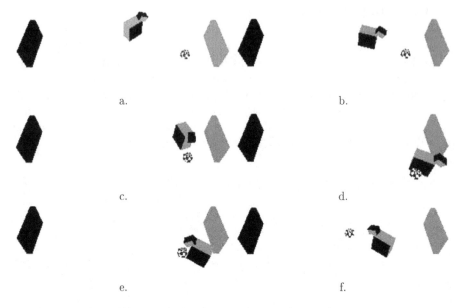

a.   b.

c.   d.

e.   f.

Fig. 6. Ball chasing sequence on the simulator.

One difference between simulation and the real world was in the collisions between the ball and the robot. In the simulator, collisions between the robot and the ball are treated like a collision between a ball and a moving box. In the real world, the robot has moving legs that cause a greater variety of results.

5   Summary

We described our simulator for AIBO. The simulator handled locomotion of the robot, sensing of the digital camera and infrared distance sensor, and physical dynamics between the objects in the world.

Using this simulator we evolved a neural-controller for ball chasing that successfully transferred to AIBO. We noticed that collisions between the robot and ball had different results in the real world than in the simulated world. This did not affect ball chasing performance but suggests that evolving more complex behaviors for AIBO to interact with a ball may require better modeling of robot-ball collisions.

References

[Fujita & Kitano, 1998] Fujita, M. & Kitano, H. (1998). Development of an autonomous quadruped robot for robot entertainment. *Autonomous Robotics*, 5:1–14.
[Gallagher & Beer, 1992] Gallagher, J. C. & Beer, R. D. (1992). A qualitative dynamical analysis of evolved locomotion controllers. In Meyer, J.-A., Roitblat, H. L., & Wilson, S. W. (Eds.), *From Animals to Animats 2*, pp. 71–80.

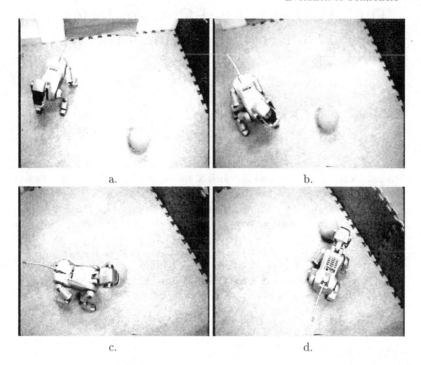

a.

b.

c.

d.

Fig. 7. Ball chasing sequence with an actual AIBO and an evolved neural network controller.

[Gallagher et al., 1996] Gallagher, J. C., Beer, R. D., Espenschied, K. S., & Quinn, R. D. (1996). Application of evolved locomotion controllers to a hexapod robot. *Robotics and Autonomous Systems*, 19(1):95–103.

[Hornby et al., 1999] Hornby, G. S., Fujita, M., Takamura, S., Yamamoto, T., & Hanagata, O. (1999). Autonomous evolution of gaits with the sony quadruped robot. In *Proceedings of the Genetic and Evolutionary Computation Conference*. Morgan Kaufmann.

[Hornby & Mirtich, 1999] Hornby, G. S. & Mirtich, B. (1999). Diffuse versus true coevolution in a physics-based world. In Banzhaf, Daida, Eiben, Garzon, Honavar, Jakiel, & Smith (Eds.), *Proceedings of the Genetic and Evolutionary Computation Conference*. Morgan Kaufmann.

[Jakobi, 1998] Jakobi, N. (1998). *Minimal Simulations for Evolutionary Robotics*. PhD thesis, School of Cognitive and Computing Sciences, University of Sussex.

[Lund & Miglino, 1996] Lund, H. H. & Miglino, O. (1996). From simulated to real robots. In *Proceedings of IEEE 3rd International Conference on Evolutionary Computation*. IEEE Press.

[Michel, 1995] Michel, O. (1995). An artificial life approach for the synthesis of autonomous agents. In Alliot, J., Lutton, E., Ronald, E., Schoenauer, M., & Snyers, D. (Eds.), *Proceedings of the European Conference on Artificial Evolution*, pp. 220–231. Springer-Verlag.

[Miglino et al., 1995] Miglino, O., Lund, H., & Nolfi, S. (1995). Evolving mobile robots in simulated and real environments. *Artificial Life*, 2(4):417–434.

The Evolution of 3-d C.A. to Perform a Collective Behavior Task

Francisco Jiménez-Morales

Departamento de Física de la Materia Condensada. Universidad de Sevilla.
P. O. Box 1065, 41080-Sevilla, Spain. jimenez@cica.es

Abstract. Here we extend previous results in which a genetic algorithm
(GA) is used to evolve three dimensional cellular automata (CA) to perform a non-trivial collective behavior (NTCB) task. Under a fitness function that is defined as an averaged area in the iterative map, the GA
discovers CA rules with quasiperiod-3(QP3) collective behavior and others with period-3. We describe the generational progression of the GA
and the synchronization necessary to maintain the global behavior is
shown using a generalized space-time diagram that reveals the existence
of propagating structures inside the system.

1 Introduction

Evolution can produce sophisticated "emergent computation" in systems composed of simple components that are limited to local interactions. The term
"emergent computation" refers to the appearance in a system's temporal behavior of information-processing capabilities that are neither explicitly represented
in the system's elementary components. In both natural and human-constructed
information-processing systems, allowing global coordination to emerge from a
decentralized collection of simple components has important potential advantages—
e.g., speed, robustness, and evolvability—as compared with explicit central control. However, it is difficult to design a collection of individual components and
their interaction in a way that will give rise to useful global information processing or "emergent computation".

In order to understand the mechanisms by which an evolutionary process can
discover methods of emergent computation a simplified framework was proposed
and studied by Crutchfield, Mitchell, and their colleagues [13, 3, 5, 4] in which
a genetic algorithm (GA) evolved one-dimensional cellular automata (CAs) to
perform computations. In their work the GA was able to discover CAs with
high performance on tasks requiring cooperative collective behavior. The density classification task and the synchronization task are two examples of emergent
computation for small radius binary CA. A successful CA for the classification
task will decide whether or not the initial configuration contains more than half
1s. If it does, the whole lattice should eventually iterate to the fixed point configuration of all cells in state 1; otherwise it should eventually iterate to the

J. Miller et al. (Eds.): ICES 2000, LNCS 1801, pp. 90–102, 2000.

fixed-point configuration of all 0s. For the synchronization task a successful CA will reach a final configuration in which all cells oscillate between all 0s and all 1s on successive time steps.

A much more complex situation of emergent behavior in CAs is found in d=3 with the appearance of non-trivial collective behavior (NTCB). As CAs are governed by local interactions and subjected to noise it was expected that any globally observable would show a trivial time dependence in the limit of infinite size [1]. But several exceptions to this have been found. The most remarkable one is a quasiperiod three behavior (QP3) that exhibits the concentration of rule-33 automaton in d=3 [9] and other CAs in high space dimensions [2]. This behavior is neither transient nor due to the finite size of the lattice and has been obtained for deterministic and probabilistic rules [10]. Several attempts have been made to understand its phenomenology and have addressed the possible mechanisms by which this puzzling collective behavior emerges [7]. But at the moment there is not any answer to the question of how NTCB can be predicted from the local rule, nor how can we design a CA with a specific behavior.

In this paper we extend previous results [11] in which we couple an evolutionary process-a GA- to a population of three dimensional CAs. The survival of an individual CA is determined by its ability to perform a "QP3(P3) task" and we describe the synchronization mechanism in terms of propagating structures in space and time.

2   Cellular Automata

Cellular Automata (CAs) are regular lattices of variables, each of which can take a finite number of values ("states") and each of which evolves in discrete time steps according to a local rule that may be deterministic or probabilistic. Physical, chemical and biological systems with many discrete elements with local interactions can be modeled using CAs. The CAs studied here are three dimensional with two possible states per cell (0 or 1) and with periodic boundary conditions. We denote the lattice size (i.e., number of cells) as $N = L^3$. A CA has a single fixed rule ϕ used to update each cell; the rule maps from the states in a neighborhood of cells to a single state $s_{i,j,k}(t)$, which is the updated value for the cell at site (i, j, k) in the neighborhood.

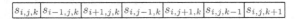

| $s_{i,j,k}$ | $s_{i-1,j,k}$ | $s_{i+1,j,k}$ | $s_{i,j-1,k}$ | $s_{i,j+1,k}$ | $s_{i,j,k-1}$ | $s_{i,j,k+1}$ |

Table 1. A bit string representing the neighborhood of a cell $s_{i,j,k}$

The lattice starts out with an initial configuration of states and this configuration changes in discrete time steps. The neighborhood of a cell at position

(i, j, k) consists of the nearest neighbors in $d = 3$ (the Von-Neumann neighborhood), and it can be displayed as a string of 7 bits. See Table 1.

Note that the number of different neighborhood configurations is $2^7 = 128$. The transition rule ϕ can be expressed as a look-up table (a "rule table"), which lists for each local neighborhood the updated state of the neighborhood's cell at position (i, j, k). A rule is a bit string that consists of 128 bits, and then the space of rules under investigation is of 2^{128} which is too large for a sequential search.

3   The QP3(P3) task

The global behavior of the CA is monitored through the concentration of activated cells at time t, $c(t) = \frac{1}{n} \sum_{i,j,k}^{n} s_{i,j,k}(t)$. According to the time series of the concentration the types of non-trivial collective behavior reported are: noisy period-1(P1), period-2 (P2), intermittent period-2 (P2i), period-3 (P3) and quasiperiod-3 (QP3). The NTCB are chaotic in the sense of Wolfram's class III; each site of the lattice follows a chaotic evolution which has no apparent relation to the global one. The global variable $c(t)$ shows fluctuations that decrease as the lattice size increases leading to a well defined thermodynamic limit.

The goal in the "QP3(P3) task" is to find a CA that starting from a random initial configuration reaches a final configuration in which the concentration oscillates among three different values, i.e. rules with P3 or QP3 collective behavior. The QP3(P3) task count as a nontrivial computation for small-radius CA because the CA needs to make computationsfor which memory requirements increases with L and in which information must be transmitted over significant space-time distances.

4   Details of Experiments

We used a genetic algorithm (GA) to evolve three-dimensional, binary state CAs to perform a "QP3(P3) task" . Our GA begins with a population of P randomly generated chromosomes listing the rule-table output bits in lexicographic order of neighborhood patterns. The fitness evaluation for each CA rule is carried out on a lattice of 10^3 cells starting from a random initial condition of concentration 0.5. After a transient time of $N/2$ we allow each rule to run for a maximum number of M iterations. The values of concentration are assembled in groups of 4 consecutive values and the fitness function $F(\phi)$ is defined by:

$$F(\phi) = \frac{4}{M} \sum_{i}^{M/4} \frac{1}{2} abs[(c_2 - c_1)(c_4 - c_2) - (c_3 - c_2)(c_3 - c_1)]_i$$

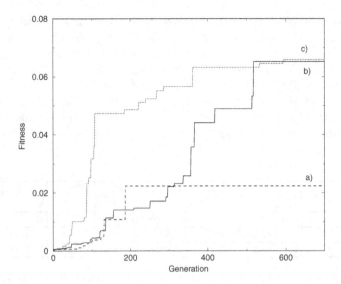

Fig. 1. Best fitness rule versus generation for three different runs. Lattice size of 10^3 cells. (a) Run in which rule ϕ_a was found at generation 188; (b) Run in which rule ϕ_b was found at generation 518; (c) Run in which rule ϕ_c was found at generation 594. The rules of the initial population were selected randomly with $0 \leq \lambda \leq 1$ in runs (a) and (b), while in run (c) $0 \leq \lambda \leq 0.5$.

The rule's fitness $F(\phi)$ is taken from a geometrical point of view and it is an average area in the iterative map, i.e. the graph of $c(t+1)$ versus $c(t)$. In the iterative map the area of a period-2 behavior is very small almost 0, the area of a noisy period-1 and the area of an intermittent P2 is higher than that of a P2 and finally QP3 and P3 behaviors have the highest values.

In each generation: (i) $F(\phi)$ is calculated for each rule ϕ in the population. (ii) The population is ranked in order of fitness. (iii) A number E of the highest fitness ("elite") rules is copied without modification to the next generation. (iv) The remaining $P - E$ rules for the next generation are formed by single-point crossover between randomly chosen pairs of elite rules. The offspring from each crossover are each mutated with a probability m. This defines one generation of the GA; it is repeated G times for one run of the GA.

5 Results

We performed more than 100 different runs of the GA with the following parameters: $M = N/2 = 500$; $P = 20$; $E = 5$; $m = 0.05$; $G = 700$ each with a different random-number seed. The dynamics of three typical runs are shown in Figure 1 which plots the fittest rule of each generation for three different runs (a),(b) and (c). Before the GA discovers high fitness rules, the fitness of the best CA rule increases in rapid jumps. Qualitatively, the rise in performance can be divided into several "epochs", each corresponding to the discovery of a new, significantly improved strategy. The best evolved rule in each run are ϕ_a , that shows a P4 behavior, ϕ_b (QP3) and ϕ_c (P3). In runs (a) and (c) the rules of the initial population had a random λ parameter in the range $[0, 1]$, while in case (b) the selected λ was in the range $[0, 0.5]$ (λ is defined as the fraction of non zero output states in the rule table).

Symbol	Rule Table Hexadecimal code	NTCB	$F(\phi)$	λ
ϕ_a	b77f3839-bb50f61a-5773f461-0d104081	P4(P2)	0.022	0.484
ϕ_b	10000000-00080c22-00020c00-80864048	QP3	0.064	0.125
ϕ_c	1000008c-0008088c-0008808b-000d0bf1	P3	0.065	0.203
R_{33}	10000008-00080886-00080886-08868621	QP3	0.053	0.164

Table 2. Measured values of $F(\phi)$, the type of non-trivial collective behavior, and λ parameter for different evolved rules: ϕ_a, ϕ_b , ϕ_c and the rule-33 automaton. Lattice size is 10^3. To recover the 128-bit string giving the output bits of the rule table, expand each hexadecimal digit to binary. The output bits are then given in lexicographic order.

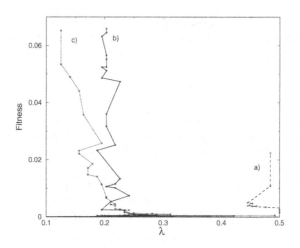

Fig. 2. Fitness function versus λ for runs (a), (b) and (c). Lattice size of 10^3 cells. Rules with high fitness values are found for $0.1 < \lambda \leq 0.2$.

Table-2 shows the rule table in hexadecimal code, the type of non-trivial collective behavior, the fitness function and the lambda parameter of the best evolved rules. Many of the CA rules that show a P2 collective behavior map low

values of concentration to high values and vice-versa. These rules have a look up
table in which there is a balance between the regions of low and high concentration values and λ is around 0.5. Rules that show QP3 or P3 behavior such as ϕ_b
and ϕ_c have a much lower value of λ. Figure 2 shows the Fitness function versus
λ for the three runs (a), (b) and (c). It has been suggested [12] that there is a
relationship between the ability of a CA rule to show complex behavior and the
λ parameter. The basic hypothesis was that λ correlates with computational capability in that rules capable of complex computation must be or are most likely
to be found near some critical value λ_c. In our experiments we have found that
rules with the highest values of F have a λ parameter in the range $0.1 < \lambda \leq 0.2$.

Usually the GA does not discover rules with QP3 or P3 behavior, but rules
with an intermittent P2 or with a P4 behavior such as the one shown in Figure 1-(a). The iterative map of the concentration for the fittest rule for different
generations in run (a) is shown in Figure 3. In the initial generations the GA discovers rules with a noisy P1 behavior, and the concentration values are around
0.5. In generation 63 there is a jump in the fitness function as an intermittent
P2 behavior is found. Another important jump in $F(\phi)$ is observed at generation 132 when a noisy P2 behavior is found. And finally in generation 188 rule
ϕ_a is discovered . From the time series of the concentration we see that rule
ϕ_a shows a P4 behavior. But in the iterative map only three clouds of points
can be distinguished: the P4 mimics a P3 and this is why the GA selects rules
like ϕ_a. When the lattice size increases the P4 collective behavior changes to a P2.

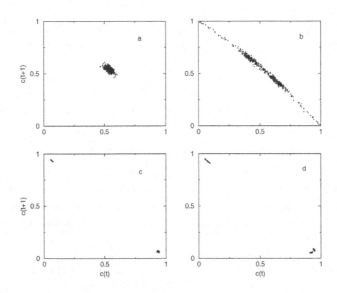

Fig. 3. Iterative map of the concentration of the fittest rule in generation: (a) 55 ; (b)
63 ; (c) 132 ; (d) 188, the best rule ϕ_a. Lattice size of 10^3 cells.

The iterative map of the concentration for the fittest rule for different generations in run (b) is shown in Figure 4. In the initial generations the GA discovers rules with a cloudy P1 behavior. In generation 100 the cloudy P1 widens and a triangular object can be seen, the GA has discovered a new rule that improves the fitness significantly . Figures 4-(c)-(d) correspond to the fittest rule in generations 156 and 518 (ϕ_b) where a QP3 can be seen clearly. In the case of a QP3, as the lattice size increases the behavior is better defined.

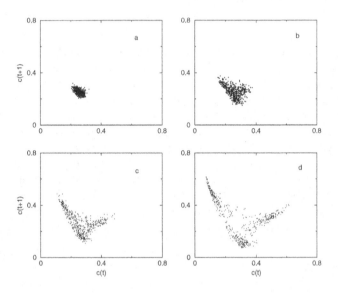

Fig. 4. Iterative map of the concentration of the fittest rule in generation: (a) 8; (b) 100 ; (c) 156 ; (d) generation 518 (ϕ_b). Lattice size of 10^3 cells.

Figure 5 shows the iterative map, the time series of the concentration and the time autocorrelation of ϕ_b. The time autocorrelation function is defined as :

$$C(t_o, t) = \frac{1}{n} \sum_{i,j,k}^{n} [s_{i,j,k}(t_o).s_{i,j,k}(t) - c(t_o).c(t)]$$

The absolute value of $C(t_o, t)$ on a log-log scale is shown in Figure 5-(c). $C(t_o, t)$ oscillates in time and the envelope of the oscillation decays as a power law with an exponent ≈ 0.5 which is according to the prediction of the Kardar-Parisi-Zhang(KPZ) equation [7].

The best way for the GA to discover rules with QP3 or P3 collective behavior is starting out with an initial population of rules with $0 \leq \lambda \leq 0.5$, such as in the run (c) of Figure 1. The iterative map of the concentration for the fittest rule for different generations is shown in Figure 6. Table-3 shows eight ancestors

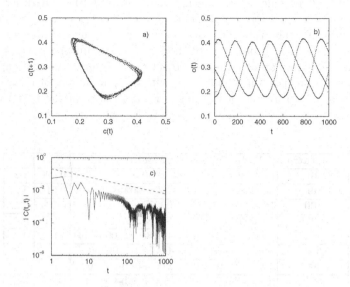

Fig. 5. QP3 collective behavior shown by rule ϕ_b, starting from a random initial concentration of 0.5. Lattice size is 10^6. Transient discarded. (a) The iterative map. (b) The time series of the concentration. (c) Log-log plot of the absolute value of time autocorrelation function. The slope of the line is $-1/2$.

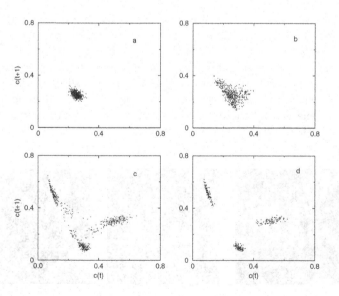

Fig. 6. Iterative map of the concentration of the fittest rule in generation: (a) 10; (b) 49 ; (c) 100 ; (d) 594 (ϕ_c). Lattice size of 10^3 cells.

of the best evolved rule ϕ_c. In the initial generations the GA discovers rules with a cloudy P1 behavior like in the previous run (b) and with a small fitness $F(\phi_{10}) = 0.0006$. The rule table of ϕ_{10} reveals that the rule maps the all 0's neighborhood to 1. In generation 49 there is a big jump in $F(\phi_{49}) = 0.0101$, and in the iterative map a triangular object can be seen. Rule ϕ_{49} maps the all 0s neighborhood to 1 and neighborhoods with a small number of 1's are mapped to 0. This seems to be one of the main characteristics of the best rules (see Table-2).

Generation	Rule Table Hexadecimal code	$F(\phi)$	λ
10	18004007-004a0868-82990002-420b6b60	0.0006	0.234
49	100008c6-004a0c0a-00088002-020b4be1	0.0101	0.219
100	1000008a-00080c0a-00088403-020b4be1	0.0317	0.203
110	10000088-00080c8a-000e8242-020d4be5	0.0464	0.226
200	10000088-0008080a-0008848b-000e0bf1	0.0487	0.195
300	10000086-0008080e-00088489-020a0bf1	0.0566	0.203
400	1000008c-00080888-0008808b-000d03f3	0.0632	0.195
550	1000008c-00080888-0008808b-000f0bf2	0.0645	0.203
594 (ϕ_c)	1000008c-0008088c-0008808b-000d0bf1	0.0659	0.203

Table 3. CA chromosomes (look-up table output bits) given in hexadecimal code, value of the fitness function and λ for eight ancestors of ϕ_c.

In generation 100 the GA discovers a rule with a QP3 collective behavior ($F(\phi_{100}) = 0.0317$). Rule ϕ_{100} maps more neighborhoods with low concentration to 0 than rule ϕ_{49}. In generation 594 the GA discovers a rule ϕ_c ($F(\phi_c) = 0.066$) which shows a P3. When the lattice size increases the P3 collective behavior changes to a QP3. Figure 7 shows three consecutive slices of the evolution of rule ϕ_c. The spatial structure is very homogeneous though there are some spot-like inhomogeneities but there is not any propagating structures such as "gliders" observed in others low-dimensional systems.

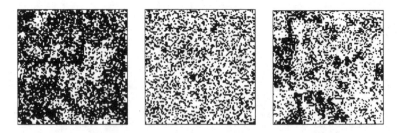

Fig. 7. Snapshots of a cross section of rule ϕ_c that shows a QP3 collective behavior running on a $d = 3$ lattice size of 10^6 cells. Three consecutive time steps are shown. Black cells correspond to state 1 and white cells to state 0.

One succesfull approach to understanding the computation performed by evolved CAs is to adopt the computational mechanics framework developed by Crutchfield and Hanson for one-dimensional CAs [8]. This framework describes the intrinsic computation embedded in the CA space-time configurations in terms of domains, particles and particles interactions. Once a CA's regular domains have been detected no-linear filters can be constructed to filter them out, leaving just the deviations from those regularities. The resulting filtered space-time diagram reveals the propagation of domain walls. If these walls remain spatially-localized over time, then they are called particles. Particles are one of the main mechanisms for carrying information over long space-time distances. In d=2 many difficulties arise when displaying the space-time diagram [6], and in d=3 the difficulties are bigger because the space-time diagram is a four dimensional surface. In order to grasp a picture of the space-time diagram we take an approximation: the space is reduced to L cells of a given axis x, and the value of concentration in each point is an average over the concentration in a two-dimensional layer perpendicular to the x axis. Figure 8(a) shows a space-time diagram of a layer-averaged concentration profile of the evolved rules ϕ_a, ϕ_b and ϕ_c. Time goes downwards and the value of concentration sets the gray scale. It can be observed regular and synchronized regions that consist of alternating values of concentration , and some other irregular ones. After the homogeneous regions have been filtered in Figure 8(b), some propagating structures can be observed. For ϕ_a the space-time diagram reveals clearly a P2 collective behavior as two alternating regions (black and white) with a superimposed noise and the filtered diagram shows a set of black points without any kind of structure. For rules with a QP3 collective behavior , such as ϕ_b and ϕ_c, the space-time diagram is more complex because the concentration values are cycling around the QP3. The coordination needed to sustain the global behavior can be seen in the filtered diagram as wave-like structures that propagate among different regions in space and time. These pictures suggest that there is some kind of a net of sites doing more or less the same thing and thus being responsible for the synchronization of the automata.

It is of interest to see how the evolutionary process such as the GA can produce sophisticated emergent computation in decentralized systems. Figure 8(c) shows space-time and filtered diagrams of a layer-averaged concentration profile, corresponding to three ancestors of ϕ_c. The cloudy P1 behaviors of the initial generations have a space-time diagram with every cell in state 0 and there are not any irregularities (this is not shown in Figure 8). When the first rule with a QP3 behavior appears in generation 100, the space-time diagram shows some structures that interact among them as can be seen in the corresponding filtered diagram. In generation 110 the irregularities between domains propagates mainly along the time. Finally in generation 200 the space-time diagram shows a wave-like structure that maintains its pattern on future generations.

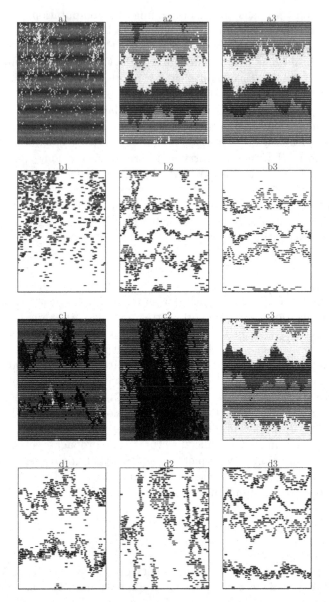

Fig. 8. (a1,a2,a3) Space-time diagram of a layer-averaged concentration profile for rules ϕ_a, ϕ_b, and ϕ_c starting from a random initial concentration. Transient discarded. Each dot represents an average over the concentration in a two-dimensional layer perpendicular to the x axis (White represents a concentration value for which $0 \leq c < 0.1$ and black when $0.4 \leq c < 0.5$). Time goes downwards and 200 time steps are shown. Lattice size of 10^6. (b1,b2,b3) Filtered space-time diagram corresponding to (a). (c1,c2,c3) Space-time diagram of a layer-averaged concentration profile for generations 100, 110 and 200 of rule ϕ_c starting from a random initial concentration. (d1,d2,d3) Filtered space-time diagram corresponding to (c).

6   Conclusion

Artificial evolutionary processes can be helpful for discovering emergent information processing capabilities in decentralized, spatially-extended models. In this paper we extended preliminary results in which we used a GA for a non-trivial collective behavior task. Under an appropriate fitness function the GA selects preferentially rules with P3 and QP3 collective behaviors. In our experiments as the GA is evolving λ parameter of the fittest rules is decreasing and the best rules are clustered in the range $0.1 < \lambda \leq 0.2$. Finally an averaged space-time diagram reveals propagating structures in the system that can explain how the global coordination arises.

Acknowledgments

Many thanks to the EvCA group of Santa Fe Institute for their assistance and suggestions.This work was partially supported by Grant No.PB 97-0741 of the Spanish Government.

References

1. C. H. Bennet, G. Grinstein, Yu He C. Jayaprakash and D. Mukamel. Stability of temporally periodic states of classical many-body systems. *Phys. Rev. A*, 41:1932–1935, 1990.
2. H. Chaté and P.Manneville. Collective behaviors in spatially extended systems with local interactions and synchronous updating. *Progress Theor. Phys.*, 87(1):1–60, 1992.
3. J. P. Crutchfield and M. Mitchell. The evolution of emergent computation. *Proceedings of the National Academy of Science U.S.A.*, 92:10742–10746, 1995.
4. R. Das, J. P. Crutchfield, M. Mitchell, and J. E. Hanson. Evolving globally synchronized cellular automata. In L. J. Eshelman, editor, *Proceedings of the Sixth International Conference on Genetic Algorithms*, pages 336–343, San Francisco, CA, 1995. Morgan Kaufmann.
5. R. Das, M. Mitchell, and J. P. Crutchfield. A genetic algorithm discovers particle-based computation in cellular automata. In Y. Davidor, H.-P. Schwefel, and R. Männer, editors, *Parallel Problem Solving from Nature—PPSN III*, volume 866, pages 344–353, Berlin, 1994. Springer-Verlag (Lecture Notes in Computer Science).
6. F. Jiménez-Morales, J.P. Crutchfield and M. Mitchell. Evolving two-dimensional cellular automata to perform density classification: A report on work in progress. In R.Serra , S.Bandini and F. Suggi Liverani, editors, *Cellular Automata: Research Towards Industry*, pages 3–14, London, 1998. Springer-Verlag.
7. H. Chaté, G. Grinstein and P. Lei-Hang Tan. Long-range correlations in systems with coherent(quasi)periodic oscillations. *Phys.Rev.Lett.*, 74:912–915, 1995.
8. J. E. Hanson and J. P. Crutchfield. Computational mechanics of cellular automata: An example. *Physica D*, 103:169–189, 1997.
9. J. Hemmingsson. A totalistic three-dimensional cellular automaton with quasiperiodic behaviour. *Physica A*, 183:225–261, 1992.
10. F. Jiménez-Morales and J. J. Luque. Collective behaviour of a probabilistic cellular automaton with two absorbing phases. *Phys. Lett. A*, 181:33–38, 1993.

11. F. Jiménez-Morales. Evolving three-dimensional cellular automata to perform a quasiperiod-3(p3) collective behavior task. *Phys. Rev. E*, (4):4934–4940, 1999.
12. C. G. Langton. Computation at the edge of chaos: Phase transitions and emergent computation. *Physica D*, 42:12–37, 1990.
13. M. Mitchell, J. P. Crutchfield, and P. T. Hraber. Evolving cellular automata to perform computations: Mechanisms and impediments. *Physica D*, 75:361 – 391, 1994.

Initial Evaluation of an Evolvable Microwave Circuit

Yuji Kasai, Hidenori Sakanashi, Masahiro Murakawa,
Shogo Kiryu, Neil Marston, and Tetsuya Higuchi

Electrotechnical Laboratory
1-1-4, Umezono, Tsukuba, Ibaraki 305-8568, Japan
kasai@etl.go.jp

Abstract. Microwave circuits are indispensable for mobile and multi-media communication. However, these circuits are very difficult to design, because of the nature of distributed-constant circuits in the microwave range (i.e., over 1 GHz). These circuits are also difficult to adjust for optimum performance, even for experienced engineers. These related problems make development costs of microwave circuits very high. In order to overcome these problems, we propose an EHW-based microwave circuit where performance adjustment is carried out automatically by a GA. This new approach of integrating a performance adjustment function within the circuit eliminates many of the design problems with associated these circuits. In this paper, we present an EHW-based image-rejection mixer circuit, which we have developed with this approach, and experimental data that demonstrates that the automatically adjusting circuit is capable of outperforming a circuit adjusted by an experienced engineer.

1 Introduction

Microwave communication plays a very important role in mobile communication and satellite broadcasting systems. Microwave circuits usually handle signals of frequencies higher than 1 GHz. However, the design and production of microwave circuits are plagued by the serious problems posed by the behavior of distributed-constant circuits in the microwave range. Parasitic capacitances and inductances in these circuits, which are extremely difficult to predict at the design stage, cause variations in the performance levels of microwave circuits.

The standard approach to this problem is to include calibration circuitry within the design of microwave circuits to adjust circuit parameters to acceptable performance levels. However, this brings new problems for circuit designers. Although some level of adjustment circuitry is essential to deal with the behavior of distributed-constant circuits, this usually involves a difficult trade-off between circuit size and complexity. Simple circuits tend to require more adjustment circuitry, which increases the overall size of the microwave circuit. On the other hand, smaller circuits with less adjustment circuitry tend to be more complex to design. This trade-off problem is accentuated by the fact that there is a shortage of experienced analog engineers capable of making the required adjustments. The consequence of all this is that the costs of microwave circuit production are very expensive, due to the low yield

J. Miller et al. (Eds.): ICES 2000, LNCS 1801, pp. 103–112, 2000.
© Springer-Verlag Berlin Heidelberg 2000

rate and the labor cost for performance adjustments.

In this paper, we propose an EHW-based [1,2] microwave circuit that is adjusted automatically by a GA [3]. Shifting the burden of performance adjustment to the GA allows designers to keep the design of their circuits simple. We also present experimental data that demonstrates that the automatically adjusting circuit is capable of outperforming a circuit adjusted by an experienced engineer. In particular, the paper describes in detail an image-rejection mixer circuit [4-7], that executes the down-conversion of microwave signals, such as satellite broadcast signals, to the intermediate frequency range. When operating, an image-rejection mixer must reject, or filter out, signals outside a target frequency band. Although experienced analog engineers can generally improve the rejection ratio to 55 dB for a circuit by manual adjustments, the EHW-based circuit has achieved a rejection ratio of more than 69 dB.

This paper is structured as follows. After a brief description of image-rejection mixers in section 2, the architecture of our proposed image-rejection mixer circuit is outlined in section 3. Following a explanation of how we have applied a evolutionary GA in this circuit in section 4, we detail an adjustment experiment for our image-rejection mixer circuit, which compared the adjustment performances for the GA, a hill-climbing method and manual adjustment. Conclusions are presented in the final section.

2 Image-rejection Mixer

Broadcast signals received by an antenna are processed by image-rejection mixers, which select the frequency for a particular TV channel and convert it to an intermediate frequency. For example, with US communication satellites, TV programs are broadcast at frequencies between 3.7 GHz to 4.2 GHz. Image-rejection mixers process incoming signals by selecting the desired signal frequency f_{si} for a TV channel with respect to a variable local oscillator (LO) frequency f_L, and converting the f_{si} to an

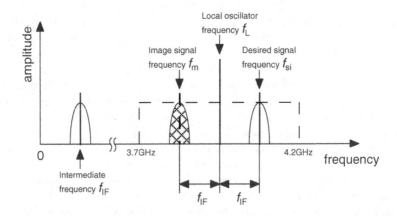

Fig. 1. Relation between desired signal frequency f_{si}, local oscillator frequency f_L, intermediate frequency f_{IF}, and image signal frequency f_m.

intermediate frequency f_{IF} (see Figure 1).

However, because the intermediate frequency f_{IF} is the difference between the desired signal frequency f_{si} and the local oscillator frequency f_L, the conversion will also include a mirror-image frequency f_m, at an equal distance from the local oscillator frequency in the opposite direction. To provide a high quality intermediate frequency signal, the mixer has to suppress this image frequency as much as possible (that is, maintain a high image rejection ratio).

Although filtering methods have been commonly used to suppress image frequencies, for these to be effective the distance between the desired frequency and the image frequency needs to be large, but this results in higher intermediate frequencies, which in turn leads to larger sized circuits. However, image-rejection mixers are able to suppress the image frequency for low intermediate frequencies.

3 An EHW-based Image-rejection Mixer

3.1 Overall Structure

Figure 2 shows the organization of an EHW-based image-rejection mixer system. It consists of two mixers, a divider, a divider/phase-shifter, and a phase-shifter/combiner.

If this circuit were set for a desired frequency f_{si} of 2.441 GHz (assumed signal bandwidth 200 kHz), with a local oscillator frequency f_L of 2.440 GHz, then the task for the image-rejection mixer would be to convert the 2.441 GHz signal to an intermediate frequency (IF) of 1 MHz, while rejecting the image frequency signal f_m of 2.439 GHz. In order to improve the image rejection ratio, the phase-shifter/combiner circuit would be evolved by GA.

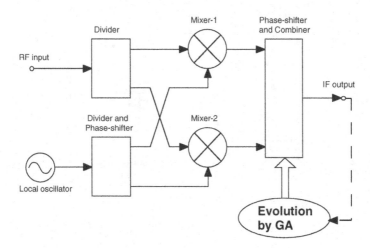

Fig. 2. Block diagram of the EHW-based image-rejection mixer.

3.2 EHW-based Image-rejection Mixer: operation and circuitry

Although the absolute performance levels of the EHW-based image-rejection mixer are not in themselves particularly remarkable, and there is undoubtedly potential to improve on the design of the mixer, which was developed using off-the-shelf discrete devices, the primary purpose of this paper is to present experimental data to show that the performance of microwave circuits can, indeed, be adjusted autonomously by GA.

The operation of the image-rejection mixer is as follows. The divider receives the microwave signals (RF signal (Radio Frequency) in Figure 2), which it feeds to the two mixers in phase. The divider/phase-shifter distributes the local oscillator signal to the mixers with a phase shift of 90 degrees. Based on the RF signal and the LO signal, the two mixers produce two IF signals which are input into the phase-shifter/combiner, which then composes an output signal with a phase shift of 90 degrees. If the values for the phase shift and the amplitude in these signal transformations match the design specification exactly, then the image frequency is canceled completely. However, due to variations in the analog devices and parasitic capacitances and inductances, it is impossible to control these phase-shift and amplitude values completely, and as a result signal quality is degraded [4,7].

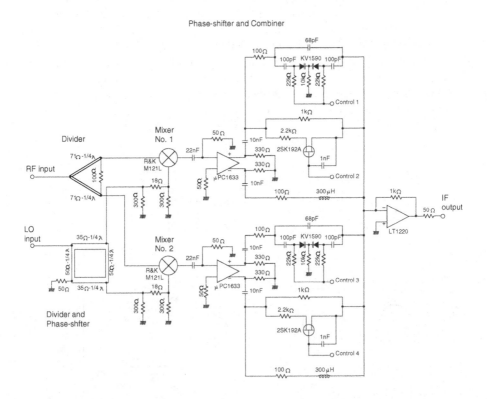

Fig. 3. Circuit diagram of the EHW-based image-rejection mixer.

The proposed image-rejection mixer, therefore, employs a GA to execute adjustments to circuitry to control the parameters of the device and to match the phase and amplitude to the exact values of a target signal. Figure 3 is a detailed circuit diagram of the image-rejection mixer. The divider and the divider/phase-shifter consist of microstrip lines (a Wilkinson coupler and a branched hybrid circuit, respectively). The two mixers (R&K M121L) both produce intermediate-frequency signals. The phase-shifter/combiner is composed of a number of discrete elements.

4 Evolution of the Image-rejection Mixer

In this section, we explain how the evolution is executed in the image-rejection mixer for automatic performance adjustments.

4.1 The Phase-shifter/combiner

A GA is executed to adjust the device parameters for the phase-shifter/combiner circuit. Figure 4 is a schematic representation of the phase-shifter/combiner. Two pairs of capacitors, resistors, and inductors shift the phase of the IF signals. By varying C1, C2, R1, and R2, the optimum phase difference and amplitude ratio between two input signals can be determined when the values of L1 and L2 are fixed.

Figure 5 shows the circuit for the capacitors C1 and C2 utilizing variable-capacitance diodes. In order to change the capacitance, a binary bit string (i.e., a chromosome for the GA) is input to a digital-to-analog converter (DAC). This represents a control voltage that is used as a reverse bias in the variable-capacitance diode. Similarly, the resistance values for the resistor circuits R1 and R2 can be controlled by the GA (Fig. 6). The output voltage of the DAC varies the resistance of the field effect transistor (FET). The lengths of the chromosomes for C1, C2, R1 and R2 are all 12 bits.

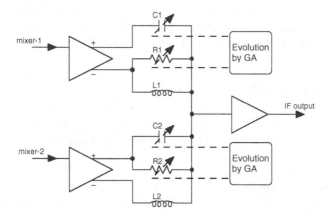

Fig. 4. Schematic representation of the adjustable phase-shifter/combiner in the image-rejection mixer.

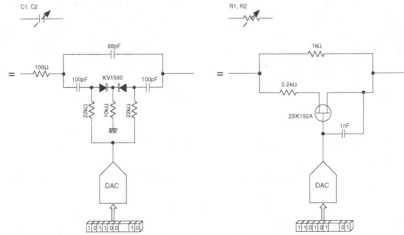

Fig. 5. Actual circuit for the adjustable capacitors C1 and C2.

Fig. 6. Actual circuit for the adjustable resistors R1 and R2.

Adjustments to C1, C2, R1, and R2 by the GA can compensate for undesirable phase shifts caused by errors in microstrip lengths and width variations, and, thus, image frequency signals can be eliminated.

4.2 Genetic algorithms

The GA operators utilized (in this application) are a uniform-crossover, with a crossover rate of 0.9, and tournament selection, with a tournament size of 2. The population size is 30, with a mutation rate of 0.05 per bit.

The fitness function for the GA is the summation of the signal intensities at three frequencies, namely, at the center and two edges of the bandwidth for a given intermediate frequency. When this summation is greater, the image rejection ratio is higher. For the adjustment comparison experiment, which we discuss in the next section, this fitness function was the summation of the intensities of three frequencies (0.9 MHz, 1.0 MHz, and 1.1 MHz) based on the intermediate frequency of 1.0 MHz.

5 Adjustment Comparison Experiment

In this section, we outline an adjustment comparison experiment conducted to evaluate the GA evolution of the image-rejection mixer against both iterated hill climbing (IHC) and manual adjustments.

5.1 Experimental Equipment

As shown in Figure 7, the experimental setup consisted of the image-rejection mixer circuit to be evolved, a local oscillator (LO), a microwave signal generator (HP

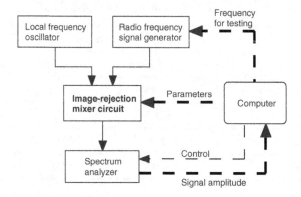

Fig. 7. Block diagram of the setup for evolving the image-rejection mixer

33120A and a frequency multiplier), a spectrum analyzer (HP 8591E with Keithley 2002), a DAC interface, and a PC. The local oscillator and the microwave signal generator were connected to the mixer input. The microwave signal generator provided the test signal (1mW) with a specified frequency according to instructions from the computer. The output signal from the mixer was monitored by the spectrum analyzer, which measured the amplitude of the output signal and transferred the values to the computer. The computer calculated the image rejection ratio and ran the GA to determine new parameters. The new evolved parameters were sent as a control voltage for the mixer circuit by the DAC.

5.2 Results and Discussion

We conducted an adjustment experiment comparing three approaches to improving the performance level of an image-rejection mixer; namely, by GA, IHC or manual adjustments by an experienced engineer. The results, which are summarized in Table

Table 1. Results of Adjustment Comparison Experiment by Adjustment Method

Searching Method	Image Rejection Ratio (dB)	
	Band Center	Band Edge
GA	69.32	38.61
IHC	41.85	35.18
Human Engineer	55	35
No Adjustment	23.40	23.38

Table 2. Table 2. Experimental Results for the IHC

	Image Rejection Ratio (dB)	
	Band Center	Band Edge
Average	41.85	35.18
Max	60.05	39.79
Min	27.49	25.43
σ^2	71.33	12.96

1, indicate that the performance after adjustment was better for the GA than for IHC and the manual adjustments

IHC repeatedly makes hill-climbing searches from different starting points selected randomly. We executed the IHC 21 times. Table 2 presents the results of the trials. The average image rejection ratio was 41.85 dB at the center of the IF band and was 35.18 dB at the edge of the IF band.

The adjustments by the human engineer were made on the same circuit. The engineer adjusted DAC values, while monitoring the spectrum analyzer display and substituting the test frequencies.

Figure 8 shows the average image rejection ratio curve for all 3 GA trials as a function of the number of generations. The image rejection ratio at the band center was 69 dB, which exceeds the best solution by the IHC search by 9 dB. At the edge of the IF band, the image rejection ratio obtained by GA was about 3 dB higher than the ratio for the IHC search. These results, therefore, indicate that the GA outperformed both the IHC method and the manual adjustments by an experienced engineer at ETL in improving the rejection ratio of the image-rejection mixer.

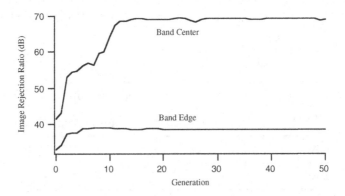

Fig. 8. Averaged image rejection ratio curve for the GA evolutions. The upper curve and the lower curve are the measured image rejection ratio at the center and at the edge of the IF band respectively.

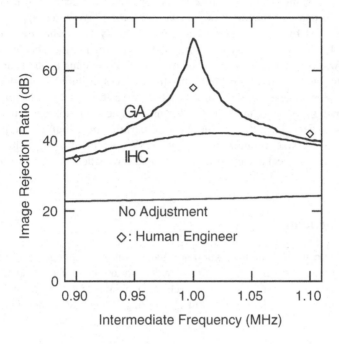

Fig. 9. Characteristics of the image rejection ratio as a function of intermediate frequency.

These results demonstrate that a GA is capable of successfully and autonomously adjusting the performance of image-rejection mixers. A further evaluation criterion for image-rejection mixers is the width of the frequency band where a high image rejection ratio is maintained. For example, as Figure 9 shows, although the image rejection ratio is at a maximum near the center of the IF band, it decreases as the frequency approaches the edge of the IF band. However, this point was not considered in the present design, being beyond the objective of this paper, which has been to present evidence that GA can adjust the performance of microwave circuits. This is something that we shall deal with in designing our next image-rejection mixer, by increasing the signal bandwidth used in the image rejection ratios.

6 Conclusions

The design and production of microwave circuits is plagued by serious, time-consuming and costly problems. Given the rapid progress in mobile and multi-media communications, there is an urgent need for innovative solutions. However, efficient methods for the design and automatic adjustments have yet to be proposed. In this paper we have proposed an EHW-based microwave circuit and presented an image-rejection mixer. The image-rejection mixer was developed in order to demonstrate that automatic performance adjustment is possible with the EHW-based approach. The present design has focused on automatic adjustment by GA rather than absolute performance levels. The results of comparison experimentation showed that its performance was better than those obtained by either hill-climbing methods or manual adjustments by human engineers. Following this successful initial evaluation of the approach, we plan to develop the image-rejection mixer further, by reducing its size and by employing a wider signal band for the image rejection ratio. Size reduction, in particular, is a very important issue for industrialization, but given previous demonstrations of size reductions by the EHW group at ETL achieved by incorporating the GA mechanisms within circuits [2], we believe, this to be a realistic goal.

Acknowledgements

We would like to thank Fumihisa Kano at Oyama National College of Technology for fabricating the mixer circuit, as well as Dr. Kazuhito Omaki, Dr. Nobuyuki Otsu, and Dr. Tsunenori Sakamoto at the Electrotechnical Laboratory for advice. We are also grateful for support from the Real World Computing Project, MITI and Dr. Junichi Shimada.

References

1. T. Higuchi, M. Iwata, D. Keymeulen, H. Sakanashi, M. Murakawa, I. Kajiani, E. Takahashi, K. Toda, M. Salami, N. Kajihara, and N. Otsu. Real-World Applications of Analog and Digital Evolvable Hardware. IEEE Transactions of Evolutionary Computation, 3 (1999) 220-235.
2. M. Murakawa, S. Yoshizawa, T. Adachi, S. Suzuki, K. Takasuka, M. Iwata, and T. Higuchi. Analogue EHW Chip for Intermediate Frequency Filters. Proceedings of International Conference on Evolvable Systems (ICES'98), (1998) 134-143.
3. J. H. Holland. *Adaptation in Natural and Artificial Systems.* The University of Michigan Press, (1975)
4. J. W. Archer, J. Granlund, and R. E. Mauzy. A Broad-Band UHF Mixer Exhibiting High Image Rejection over a Multidicade Baseband Frequency Range. IEEE Journal of Solid-State Circuits, vol. SC-16 (1981) 385-392.
5. A. Minakawa and T. Tokumitsu. A 3-7 GHz Wide-Band Monolithic Image-Rejection Mixer on a Single-Chip. IEICE Trans. Electron, E-76-C (1993) 955-960.
6. W. Baumberger. A Single-Chip Image Rejecting Receiver for the 2.44 GHz Band Using Commercial GaAs-MESFET-Technology. IEEE Journal of Solid-State Circuits, 29 (1994) 1244-1249.
7. J. C. Rudell, J. J. Ou, T. B. Cho, G. Chien, F. B. Rianti, J. A. Weldon, and P. R. Gray. A 1.9 GHz Wide-Band IF Double Conversion CMOS Receiver for Cordledd Applications. IEEE Journal of Solid-State Circuits, 32 (1997) 2071-2088.

Towards an Artificial Pinna
for a Narrow-Band Biomimetic Sonarhead

DaeEun Kim, Jose M. Carmena* and John C.T. Hallam

Institute of Perception, Action and Behaviour
Division of Informatics, University of Edinburgh
5 Forrest Hill, EH1 2QL Edinburgh, Scotland, UK
{daeeun,jose}@dai.ed.ac.uk

Abstract. A genetic algorithm was used to evolve bat-like pinna shapes for a biomimetic sonarhead. Pinnae model consisted of small reflectors placed around the transducer. Experiments with ten reflectors showed the problem of phase cancellation in the received echoes. Analysis of phase cancellation suggests more realistic pinna models for future developments.

1 Introduction

Bats are very dynamic creatures; while flying they move their wings, head, pinnae and the nose or mouth whenever they emit. They can be divided into two broad non-taxonomic groups: broadband echolocators, or fm-bats, such as *Myotis lucifugus*, whose cry consists of a frequency-swept from around 30-90 kHz; and narrowband echolocators, or cf-bats, who emit a call where about all the energy is in a single tone (for example 83 kHz for the *Rhinolophus ferrumequinum*).

Narrowband echolocators use pinna[1] motion to alter the directional sensitivity of their perceptual whereas broadband listening systems (*e.g.* humans and broadband emitting bats) rely on pinna *morphology* to alter acoustic directionality at different frequencies [9]. The importance of pinna motion along vertical arcs in the cf-bat for target localization in the vertical plane has been investigated with real bats [1, 2, 7]. The use of this motion might be the reason that the *Rhinolophus ferrumequinum* has unusually large pinnae compared to the size of its head as can be seen in figure 1.

The relationship between bats and robots arises because the sensor interpretation problems of bats while navigating in cluttered environments such as forests are very similar to those of mobile robots provided with ultrasonic sensors when navigating in laboratories. Moreover, the constant frequency pulse emitted by the cf-bat when echolocating is analogous to the one emitted by robotic ultrasonic sensors in terms of bandwidth. For their experiments, Walker *et al.* [9] used a robotic model composed of a 6 degree of freedom biomimetic sonarhead (figure

---

* J.M. Carmena is supported by the European Union TMR Network SMART2 under contract number FMRX-CT96-0052.
[1] The complex convoluted external ear.

J. Miller et al. (Eds.): ICES 2000, LNCS 1801, pp. 113–122, 2000.

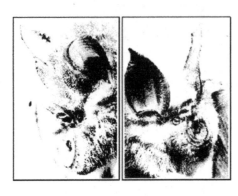

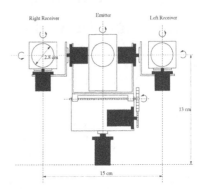

Fig. 1. *Rhinolophus ferrumequinum* (Photos from [3]).

Fig. 2. Diagram of the biomimetic sonarhead.

2) mounted on a mobile robot [5]. In our work we are interested in integrating the cf-bat's sensorimotor system for obstacle avoidance in this robotic model (exploiting the physical capabilities of the sonarhead), as a biological approach to ultrasonic-based navigation in mobile robots.

Thus, as part of our working plan, we want to improve the directional sensitivity of the sonarhead's receivers (*i.e.* maximise the angular resolution of the receiving transducers) by adding artificial pinnae to them. Because of the difficulty of designing a pinna model by an analytical approach, an evolutionary approach consisting of a GA together with a software model of the sonarhead [8], in which to evaluate the evolved solutions, is used instead. Our work continues that of Papadopoulos, and of Peremans *et al.* [4,6], who used genetic algorithms to evolve simple pinna shapes for broadband echolocators.

1.1 Narrow-band 3D target localisation

It is quite interesting to see the way in which echolocators with narrow-band call structures perform target localisation. In the case of the cf-bat, this localisation is performed mostly using the information contained in a single harmonic echo. In order to calculate the target's azimuth angle with a receiver placed on each side of the head (as in bats), interaural intensity differences (IIDs) as well as interaural time differences (ITDs) can be employed.

However, how can the elevation angle be estimated? Experiments with the biomimetic sonarhead [9] showed how, by sweeping a pair of receivers through opposite vertical arcs (figure 3), dynamic cues, in the form of amplitude modulations which vary systematically with target elevation, are created (figure 4). Thus, by this arc scanning, a delay-per-degree transformation is created.

This, combined with the azimuth angle estimation by means of IIDs and the target's range by the echo delay, provides a narrow-band echolocator with a 3D estimation of an insonified target's relative position.

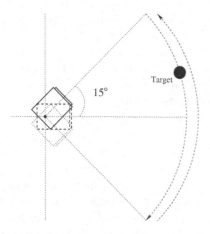

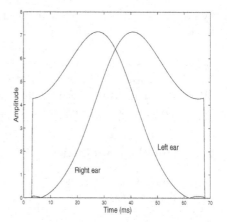

Fig. 3. Side view of sonarhead's receiver performing vertical arcs for elevation angle estimation.

Fig. 4. Delay between amplitude peaks encodes target elevation (target elevation at 6 deg.).

2 Artificial pinnae

2.1 Previous work

Previous work in evolving pinna morphology [4, 6] focussed on broadband echolocators. The pinna was modelled by up to three disc reflectors whose position and orientation angle around the receiving transducer were determined by a genetic algorithm (GA), obtaining a chromosome with the following structure,

$$(x_1 \ y_1 \ z_1 \ \alpha_1 \ \beta_1 \quad x_2 \ y_2 \ z_2 \ \alpha_2 \ \beta_2 \quad \ldots \quad x_n \ y_n \ z_n \ \alpha_n \ \beta_n)$$

where x,y and z are cartesian position coordinates and α, β are azimuth and elevation angles. The GA comprised a population of candidate sets of reflector positions, whose fitness was determined by simulating their effect on the acoustic signals transduced by the receiver. *2-point* crossover and a mutation rate of 0.03 were used with a population of 100. A tournament selection scheme wherein a set of genomes is randomly selected from the population was used. The fittest genome was selected with a given probability; if it is not selected, then the second best is selected with the same probability, and so on. Experiments were run for 1000 generations [6].

The GA in [4, 6] was set for two tasks: first, to deploy reflectors in a monaural system so as to maximise the displacement between the axes of maximal sensitivity at 30 kHz and 90 kHz (thereby allowing target elevation to be most accurately inferred from the different amplitudes of the echo at these frequencies); and second, to deploy reflectors in a binaural system to produce a maximally steep IID curve with respect to target angular position (thereby maximising the angular resolution of the binaural system and allowing the target's position to be most accurately estimated from the IID). In the binaural case, the left ear was symmetrical with the right ear, i.e. the two pinna configurations were derived from the single disposition of reflectors indicated by the GA.

The results for the first experiment were reasonable, but for the second experiment no significant improvement of the IID performance could be obtained with up to three reflectors. We therefore began (see section 2.3) by repeating the IID experiment from [6], then changing the model to the narrowband (cf-bat) case, in each case allowing up to 10 reflectors to be used by the GA.

2.2   Model considerations

In this work we use similar model considerations as [6], that is: disc-shaped specular reflectors are used to modify the directionality characteristics of a dynamic binaural echolocation system. The differences with respect to [6] are the consideration of sound losses in the reflectors due to absorption, instead of considering perfect reflection, and the way in which we calculated the phase cancellation phenomena. We assumed an absorption rate of the reflectors of 20% of the incident sound. Phase cancellation among different echos from the reflectors when arriving at the transducer is also considered.

As in [6], the diffraction and diffusion phenomena around the edges of the reflector discs are considered insignificant. Also, no multiple reflections are taken into account, i.e. each reflector introduces one additional echo path. The reflectors' radii are constant and equal to that of the receiver. The reflector orientation angles vary between -90 and +90 degrees with a resolution of 2 degrees.

2.3   Results

The first step consisted of the repetition of the experiments reported in [6]. Figure 5 shows the directivity differences between a bare transducer and a transducer with 3 reflectors, and IID changes at various elevation angles. These results matched with the results reported in [6] so we decided to continue scaling up the model for more reflectors.

3   The evolutionary approach

Our goal is to evolve a reflector formation around the transducers for a desirable pinna shape for improving IID and arc scanning behaviours. The reason for taking this reflector approach instead of a more complex one, such as surface formation, is because of the extensive computational time the latter would take.

3.1   Methods

Two different methods were considered, a signal based method and a region coverage method, the latter being the chosen one because of its smaller processing time.

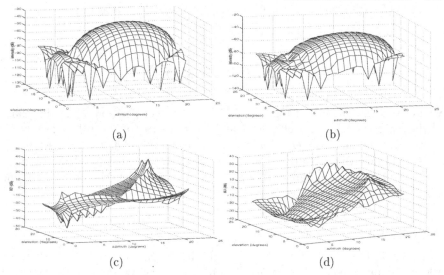

Fig. 5. Comparison between bare transducer and transducer with reflectors (a) directivity with bare transducer (b) directivity with reflectors (c) IID with bare transducer (d) IID with reflectors

Signal based method This method is based on the one used in [6] for IID behaviour. In our case, because of the arc scanning behaviour, we were seeking a high amplitude-modulated signal with sharp peaks for a better delay-per-degree estimation. For that purpose, a fitness function $F = A/\sigma$, where A is the maximum amplitude value during the arc scanning and σ is the standard deviation of the time-varying amplitude along time during the arc scanning, was used. This fitness function would possibly guarantee clarity of target position. The method was finally rejected because of the high amount of computational time required.

Region coverage method This is based on the following assumption: having an ear morphology whose left ear focuses on the left side of the target's position along azimuth angle and the right ear on the right side, a broader range of IIDs can be obtained. In this method we also sought to evolve a reflector formation for both a good IID range and arc scanning. For the IID case, targets at every azimuth and elevation angle were considered while for the arc scanning case we only considered *slices* of the vertical plane, *i.e.* all elevation positions for a fixed azimuth.

3.2 Fitness function

The fitness function aimed to combine the covered region method with the phase cancellation constraint, thus no reflector should be positioned in a location where, from any of the possible target positions, phase cancellation happens.

Based on this criterion, our fitness function was defined as

$$L = \alpha \mid \sum_{k=0}^{N} e^{-i(wt+\theta_k)} \mid + \beta(\sum_{i=1}^{N} \sum_{k=1}^{M} r_{ik})$$

where N is the number of reflectors, M is the target position, θ_k is the phase of the wave coming from reflector k and r_{ik} is set to 1 if the i-th reflector can reflect the left-sided target k on the left transducer (otherwise it is set to 0). This will allow the reflectors around one transducer to focus on the side of it. As a result, it should improve IID range by increasing the echo intensity of one ear with respect to the other.

3.3   Results

When using the above fitness function, results for IID using 10 reflectors were very little improved from those in [6]. In figure 6, a reflector distribution around the transducer (a) and the region covered by these reflectors (b) is shown.

As can be seen, the middle part of the IID profiles, $i.e.$ the part related to the main lobe of the transducer directivity, is quite similar to the bare transducer and to the 3 reflector case (figure 6(c)) in terms of steepness and linearity and therefore there is no improvement. However, there is a small improvement in the side lobe parts of the IID profile, for a target at 2 deg. elevation angle, in the form of smoothness of the peaks of such side lobes. A smoother performance along these lobes ($i.e.$ removing the peaks) offers an improvement of the angular range along the horizontal plane. Arc scanning behaviour with this reflector configuration (d) performs fairly well, that is, there is some distortion in the wave peaks (continuous line) which is the significant part for arc scanning, but this could be resolved by a suitable curve-fitting process, $e.g.$ using the bare transducer curve (dotted line).

When evolving a reflector configuration for arc scanning behaviour (figure 7), results are slightly more satisfactory than in figure 6(d)). In figure 7(b), there is an improvement in amplitude (continuous line), despite some distortion at the middle part of the scan, compared with bare transducer (dotted line).

From this results, it is clear that there are no big improvements in a 10 reflector configuration with respect to [6]. In order to investigate the reasons for this performance we analysed the simplest reflector case, as seen next.

4   Phase matters

The reason there is no big improvement in performance is the effect that phase cancellation produces in the final wave. Because of the difficulty of finding an optimal position for all the reflectors in all the target possible positions, final performance does not significantly improve with respect to a bare transducer configuration, as the analysis below suggests.

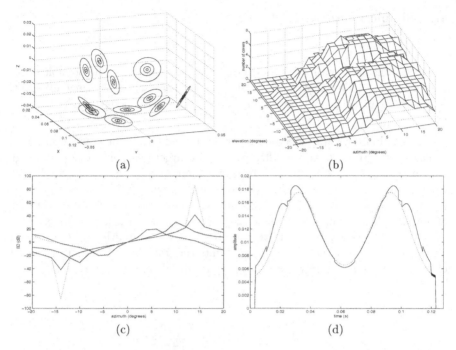

Fig. 6. IID results: (a) reflector formation (b) region covered by reflectors (c) IID at elevation 2, 13, and 18 degrees (d) arc scanning at azimuth 0 degree (dotted: bare transducer; solid: transducer with reflectors).

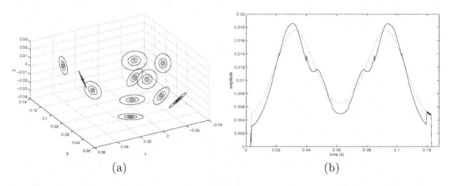

Fig. 7. Arc scanning results: (a) reflector formation (b) arc scanning at azimuth 0 degree (dotted: bare transducer; solid: transducer with reflectors).

For investigating the effect of phase cancellation, we used a single reflector configuration and then we scaled it up to three reflectors. In both cases, the transducers were considered static, *i.e.* no arc scanning behaviour was considered.

4.1   Simplest case: one reflector

A static transducer with a fixed reflector was evaluated for all the possible target positions, that is, an array of 21 × 21 positions representing a range of −20 to +20 degrees in both horizontal and vertical planes (the sonarhead's resolution is 2 deg.).

As seen in figure 8(a), the reflector is positioned beside the transducer on the right ear. The reflector can be effective only in target positions at azimuth angle ranging from -20 to -14. Figure 8(b) shows the region covered by this reflector; each cell shows a phase diagram of how much signal of the reflector's echo is phase-shifted from the transducers's direct echo signal. The reflected signal is about 235 degree phase-shifted at -20 degrees in azimuth and -18 degrees in elevation (c), and about 180 degrees at -16 degrees in azimuth and -6 degrees in elevation (d). The net effect of the signals into the transducer is the superposition of all the incoming signals and we can see the result of echo interference.

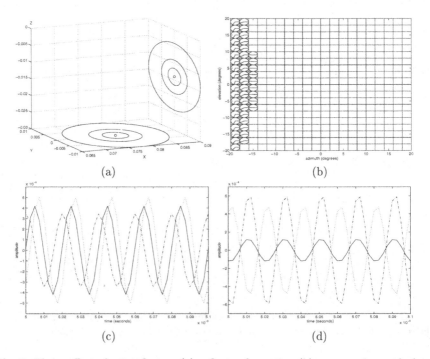

Fig. 8. Phase effects for 1 reflector: (a) reflector formation (b) cover region and phase shift (c) echo signals at azimuth -20, elevation -18 (d) echo signals at azimuth -16, elevation -6 (dotted: transducer; dashed: reflector; solid: superposition of signals).

4.2   Three reflectors case

In this case a three reflection configuration was used (figure 9(a)). The regions coverage of each reflector (b), are located on the right side as expected. The echo interference becomes more complex when we have more reflectors. As a matter of fact, echo interference occurs with a large number of signals from any surface around the transducer position. In some cases the net echo signal is overwhelmed by the direct signal to the transducer (d), and in other cases the reflected signals greatly influence the net echo becoming unpredictable for a target position. Thus, it is not possible to focus on all target positions with a good reflector formation. Even a slight movement of target position makes a phase shift signal for one reflector as in 8(b). This results in our objective being a very difficult problem to solve with a genetic algorithm.

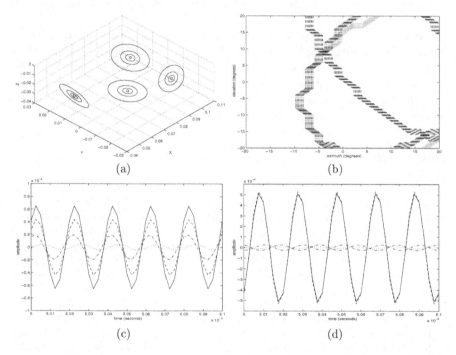

Fig. 9. Phase effects for 3 reflectors (a) reflector formation (b) contour of cover region (c) echo signals at azimuth -2, elevation -20 (d) echo signals at azimuth 20, elevation 14 (dash: transducer; dashdot: reflector 1 and 3; dot: reflector 2; solid: superposition of signals).

5   Discussion and further work

This work aimed to improve the results obtained in [6] in which artificial pinnae in the form of three reflectors were attached to a biomimetic sonarhead. The goal was to improve on that work for IID behaviour and arc scanning. Using a GA similar to [6], a region coverage method was used to evolve pinna shapes of up to ten reflectors. From the results obtained, we realised that increasing the number of reflectors from three to ten does not improve performance enough, because of the adverse effect of multipath *phase cancellation* phenomena. Experiments with one and three reflector models showed how the effect of phase cancellation for a fixed reflector configuration varies for different target positions. Our conclusion from these results is that we were using too simple a model of the pinnae, *i.e.* using small reflectors instead of surfaces. Hence, to evolve an optimal reflector configuration which will improve performance for every target position using this simple model seems to be a very difficult task.

This suggests that a more realistic model of the bat's pinna would be a fruitful avenue to explore. However, the problem with a more realistic surface-based model is the substantial increase in parameters required and hence in the space to be searched by the GA.

At this point, we propose as further work to investigate good compromises for the tradeoff mentioned above. As an example, we propose to use parabolic surfaces in which many small reflectors would be placed around the focus point. Because of the inherent properties of the parabola equation, all the reflections will direct to the focus, *i.e.* the transducer. Also, from the intersection of different parabolic surfaces we might obtain better results.

References

1. D.R. Griffin, D.C. Dunning, D.A. Cahlander, and F.A. Webster. Correlated orientation sounds and ear movements of horseshoe bats. *Nature*, 196:1185–1186, 1962.
2. J. Mogdans, J. Ostwald, and H.-U. Schnitzler. The role of pinna movement for the localization of vertical and horizontal wire obstacles in the greater horshoe bat, *rhinolophus ferrumequinum*. *J. Acoust. Soc. Amer.*, 84(5):1988, November 1988.
3. R.M. Nowak. *Walker's Bats of the World*. Johns Hopkins University Press, 1994.
4. G. Papadopoulos. Evolving ears for echolocation. Master's thesis, Department of Artificial Intelligence, University of Edinburgh, 1997.
5. H. Peremans, A. Walker, and J.C.T. Hallam. A biologically inspired sonarhead. Technical Report 44, Dep. of Artificial Intelligence, U. of Edinburgh, 1997.
6. H. Peremans, V.A. Walker, G. Papadopoulos, and J.C.T. Hallam. Evolving bat-like pinnae for target localisation by an echolocator. In *Proceedings of the Second International Conference on Evolvable Systems: From Biology to Hardware*, pages 230–239. 1998.
7. J.D. Pye and L.H. Roberts. Ear movements in a hipposiderid bat. *Nature*, 225:285–286, 1970.
8. V. A. Walker. *One tone, two ears, three dimensions: An investigation of qualitative echolocation strategies in synthetic bats and real robots*. PhD thesis, University of Edinburgh, 1997.
9. V. A. Walker, H. Peremans, and J. C. T. Hallam. One tone, two ears, three dimensions: A robotic investigation of pinnae movements used by rhinolophid and hipposiderid bats. *J. Acoust. Soc. Amer.*, 104(1):569–579, July 1998.

Towards a Silicon Primordial Soup:
A Fast Approach to Hardware Evolution
with a VLSI Transistor Array

Jörg Langeheine, Simon Fölling, Karlheinz Meier, Johannes Schemmel

Adress of principle author: Heidelberg University, Kirchhoff-Institut für Physik,
Schröderstr. 90, D-69120 Heidelberg, Germany, ph.: ++49 6221 54 4359
langehei@kip.uni-heidelberg.de

Abstract. A new system for research on hardware evolution of analog
VLSI circuits is proposed. The heart of the system is a CMOS chip providing an array of 16 × 16 transistors programmable in their channel
dimensions as well as in their connectivity. A genetic algorithm is executed on a PC connected to one or more programmable transistor arrays
(PTA). Individuals are represented by a given configuration of the PTA.
The fitness of each individual is determined by measuring the output of
the PTA chip, yielding a high test rate per indiviuum. The feasibility of
the chosen approach is discussed as well as some of the advantages and
limitations inherent to the system by means of simulation results.

1 Introduction

Analog circuits can be much more effective in terms of used silicon area and
power consumption than their digital counterparts, but suffer from device imperfections during the fabrication process limiting their precision ([1]). Furthermore
any progress in design automation seems to be much harder to achieve for analog
circuits than for digital ones. Recently the technique of genetic algorithms (GA)
has been applied to the problem of analog design with some promising results.
The capability of artifical evolution to exploit the device physics available has
been demonstrated for example in [2].

The various attempts to the evolution of analog circuits range from purely extrinsic evolution that simulates the circuits composed by the genetic algorithm
(GA) (see e.g. [4]) over breeding of analog circuitry on FPGA's designed for
digital applications ([2], [3]) and the use of external transistors (e.g. [5]) as well
as the optimisation of parameters of an otherwise human designed circuit ([6])
to the use of chips designed with certain design principles in mind ([7]).

However, one of the most elementary devices for analog design is given by the
transistor, which can also be used to form resistors and capacitors. Therefore
the development of a hardware evolution system employing a programmable
transistor array (PTA) to carry out the fitness evaluation intrinsically in silicon
is proposed.

J. Miller et al. (Eds.): ICES 2000, LNCS 1801, pp. 123–132, 2000.

The evolutionary system aimed at will hopefully yield the following advantages compared to the approaches described above: Except for dc-analyses the evaluation of the fitness of the according individual can be carried out much faster than in simulation, hopefully yielding a cycle time of less than 10 ms. In contrast to simulations the GA has to deal with all the imperfections of the actual dice produced and thus can be used to produce circuits robust against these imperfections (a discussion on the evolution of robust electronics can be found in [3]). Furthermore, circuits evolved on a custom made ASIC can be analyzed more easily than on a commercial FPGA, since in general the encoding of the loaded bitstrings is not documented for commercial FPGAs. On the contrary for custom made ASICs fairly well suited models can be used to simulate the circuit favoured by the GA. Analysis of the attained circuits will be further enhanced by the integration of additional circuitry allowing to monitor voltages and currents in the transistor array itself as well as the die temperature. Finally, a lot of the structures designed for the PTA could be reused to set up a whole family of chips that merely differ in the elementary devices used for evolution. The programmable transistors could be replaced, for example, by programmable transconductance amplifiers, silicon neurons, or operational amplifiers combined with a switched capacitor network. Thereby different signal encodings (voltages, currents, charge, pulse frequencies and so on) could be investigated and compared with regard to their 'evolvability'.

The rest of the paper is organized in a bottom up fashion: Section 2 describes the structure of the programmable transistor cell. In section 3 the PTA chip as a whole is discussed, while section 4 presents the embedding of the chip in a system performing the artifical evolution. Finally in section 5 some simulation results will be given demonstrating the feasibility as well as the limits of the proposed approach, before the paper closes with a summary.

2   Design of the basic transistor cell

2.1   Choosing the transistor dimensions and types

The core of the proposed chip consists of an array of 16×16 programmable transistor cells, each acting as one transistor in the programmed circuit. Each cell contains an array of 20 transistors providing 5 different lengths and 4 different widths. The 5 different lengths are decoded by three bits and vary logarithmically from $0.6\,\mu$m to $8\,\mu$m. Since a transistor of a given width W can be approximated fairly well by using two transistors of the same length and width $W_1 + W_2 = W$ in parallel, 15 different widths can be chosen ranging from 1 to $15\,\mu$m. Relatively small steps between adjacent lengths and widths are chosen to attain a fitness landscape smooth with regard to variations of the channel dimensions of the used transistors. A rather large number of different lengths is chosen, because the characteristics of a MOS transistor do not only depend on the aspect ratio (W/L), but also on its actual length (cf. [8]).

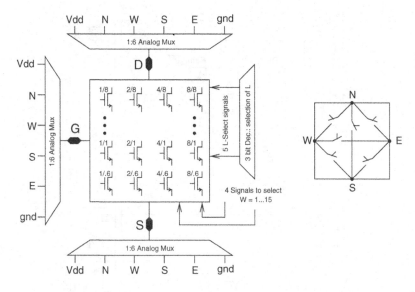

Fig. 1. Basic NMOS cell: **Left:** The cell contains 4 × 5 different transistors. **Right:** Routing capability of one cell.

2.2 Connecting the chosen transistor

The chosen transistors are connected to the three terminals D (drain), G (gate) and S (source) through transmission gates (Fig. 1 shows an NMOS cell. Since the process used is based on p-doped silicon, there is an additional terminal B for the bulk connection of the PMOS cell). Each of these 'global' transistor terminals can be connected to either of the 6 terminals listed at the output of the 1:6 analog multiplexer. The terminals represent connections to the four adjacent transistor cells, power and ground. In this architecture the signal has to pass two transmission gates from the actually chosen transistor to the outside of the cell, resulting in twice the on-resistance of one transmission gate. On the other hand the two-fold switching saves a lot of transmission gates (20 + 6 instead of 20 × 6 for each 'global' terminal) and thereby a lot of parasitic capacitance.

On the right hand side of Fig. 1 the routing capability of the cell is shown. Signals can be connected from any of the four edges of the cell to any of the three remaining edges via one transmission gate. In total

$$4 \ (W) + 3 \ (L) + 3 * 3 \ (\text{Multiplexing of D,G,S}) + 6 \ (\text{Routing}) = 22 \qquad (1)$$

bits are needed for the configuration of one NMOS cell. In case of the PMOS cell another three bits must be added to multiplex the bulk terminal.

3   The PTA chip

For the chip an array is formed out of 16 × 16 programmable transistor cells as indicated in Fig. 2. PMOS and NMOS transistors are placed alternatingly

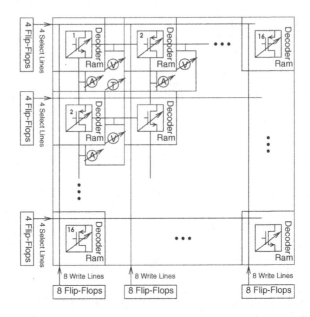

Fig. 2. Array of programmable transistor cells including S-RAM cells for configuration storage and readout electronics allowing to monitor voltages and currents inside the network as well as the die temperature. The digital parts of the array are shaded in darker gray than the analog ones. PMOS cells are distinguished from the NMOS cells by their white shading.

resulting in a checkerboard pattern. With 256 transistors in total, circuits of fairly high complexity should be evolvable.

3.1   Configuration of the programmable transistor cells

The configuration bits for determining the W and L values and the routing of each cell are stored in 32 static RAM bits[1] per cell, yielding $256 * 32 = 11392$ bits for the configuration of the whole chip. In order to test individuals with a frequency of $100\,Hz$, the configuration time should be small compared to a cycle time of $10\,ms$, e.g. $100\,\mu s$. Writing 8 bits at a time as shown in Fig.2 thus results in a writing frequency of about $14\,MHz$, which is easily achieved with the $0.6\,\mu m$ process used. As shown in Fig. 3 the actual configuration can be read back, which is useful to control the writing of the configuration data as well as their stability in the possibly noisy environment. The digital parts of each cell are laid out around the analog parts to allow fast replacement of the core part of the cell enabling the testing of different 'elementary' devices as already discussed in section 1.

---

[1] The choice of 32 instead of the needed 25 is taken with regard to future implementations of different corecells into the same surrounding infrastructure.

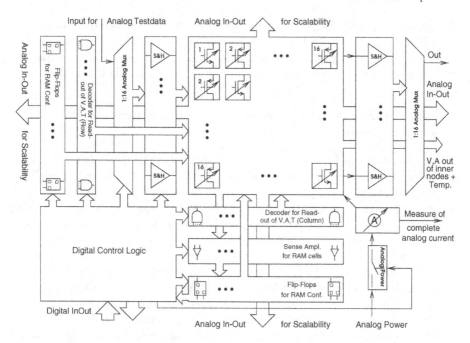

Fig. 3. Block diagram of the PTA chip. The analog multiplexer and the sample and hold stages for in- and output of the border cells are shown on one side of the transistor array only for simplicity. Actually all 64 border cells can be read out and written to.

In order to allow all possible configurations to be loaded into the PTA without causing the chip to destroy itself, two precautions are taken: Firstly all metal lines are made wide enough to withstand the highest possible currents expected for this particular connection. Secondly the analog power for the transistor array can be switched off, if the die temperature exceeds a certain limit.

3.2   Analog in- and output signals

The input for the analog test signals (meant to be voltages) can be multiplexed to the inputs of any of the 16 cells of any of the four edges of the array (For simplicity this is shown for one edge only in Fig. 3). The signal for each input is therefore maintained by means of a sample and hold unit. Similarly all of the 64 outputs at the border of the array are buffered by sample and hold circuits that can be multiplexed to the analog output (Again only shown for one edge in Fig. 3). That way the in- and output(s) of the circuit being evolved can be chosen freely, such that different areas of the chip can be used for the same experiment without loss of symmetry for the signal paths.

The output amplifiers are designed to have a bandwidth higher than 10 MHz, which should be sufficient regarding the expected bandwidth of the transistor array (cf. section 5).

All 64 outputs can be accessed directly via bond pads (referred to as 'Analog In-Out for Scalability' in Fig. 3). That way the outputs can be connected to external loads, and experiments using more than 256 programmable transistor cells can be set up by directly connecting an array of dice via bond wires.

3.3 Monitoring of node voltages, intercellular currents and temperature

As already mentioned in section 1 the PTA chip offers the possibility to read out the voltages of all intercellular nodes as well as the currents flowing through the interconnection of two adjacent cells. While the former one can be read out with several MHz, current measurements are limited to bandwidths much smaller than the bandwidth of the transisor array in order to limit the area occupied by measurement circuitry to a reasonable percentage of the chip area. Since one cannot (and maybe does not even want to) prevent the GA from evolving circuits wasting power or oscillating, the environment in the PTA may be very noisy. Accordingly the node voltages have to be buffered as closely to their origin as possible. For the same reason the current and temperature signals are locally amplified and transformed into differential currents that are multiplexed to the edges of the transistor array, where they are transformed into voltages mapped to the according pads (Fig. 3 indicates the multiplexing).

The power net for the transistor array is separated from the analog power of the rest of the chip to be able to measure the overall current consumption as well as to provide different supply voltages allowing for example the evolution of low voltage electronics.

4 The system around the chip

The architecture of the evolution system is shown in Fig. 4: The GA software

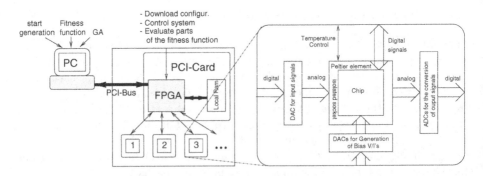

Fig. 4. Architecture of the system: **Left:** Motherboard housing several daughterboards connected to the PC via a PCI card. **Right:** One of the daughterboards presented in closer detail.

is executed on a commercial PC and communicates with the PTA chip using an FPGA on a PCI card. The FPGA gathers PTA output data and does basic calculations, the software then uses the preprocessed data to calculate the circuit's fitness. The PTA board and the software operate asynchronously to minimize cycle time. For further increase of the evaluation frequency the software is designed for maximum scalability in terms of additional processors, boards or additional computer/board systems. The FPGA distributes the digital signals for the DACs and the ADCs to the daughterboards shown on the right side of Fig. 4. A daughterboard contains a DAC for the evaluation input signal generation, DAC's providing some bias voltages and currents and ADCs for the conversion of the analog output signals of the chip (i.e. the output of the transistor array, the measured total current consumption, the signals representing die temperature and intercellular node voltages and currents.). Furthermore, a temperature control via a Peltier element will be implemented to control the temperature of the chip.

As was already pointed out by [3], the usefulness of evolved circuits strongly depends on their robustness against variations of the environment experienced by the chip. In order to provide a possibility for the evolution of robust electronics the system will be capable of testing the same individuals in parallel under different conditions, e.g. different temperatures or different dice (maybe even wafers).

5   Simulation results

As already mentioned in section 2 parasitic resistance as well as parasitic capacitance is introduced with every switch used for choosing a special transistor

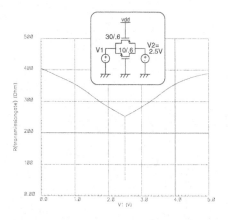
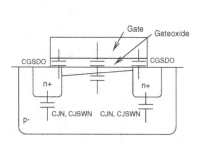

Fig. 5. Parasitic properties of the transmission gates **Left:** Simulated resistance (setup shown in the inset) **Right:** Two dimensional cut through a MOS transistor indicating its different capacitances.

geometry or routing possibility. Since increasing the switch size to obtain lower resistance values will increase its capacitance, a tradeoff has to be found. The on-resistance of a switch realized as a transmission gate with one node connected to 2.5 V is shown on the left side of Fig. 5, together with the setup for the according simulation (The glitch at 2.5 V is due to the limited computational accuracy.). For the chosen transistor geometries the on resistance is of the order of 300 to 400 Ω.

5.1   Resistance and capacitance of the switches used

The right part of Fig. 5 shows the cross section of an NMOS transistor and its capacitances. For an open switch capacitances include the gate source/drain overlap as well as the n^+ p^- capacitance per area (CJN) and per width of the transistor (CJSWN). For the transmission gate simulated in the left part of Fig. 5 these add up to 54 fF. For the multiplexing of one 'global' D,G,S or B terminal (cf. Fig. 1) the 7 open switches result in a node capacitance of 0.38 pF, while for the 31 open switches for multiplexing the D,G,S and B terminals of the actually chosen transistor inside the cell to the global terminals a node capacitance of 1.7 pF is obtained.

5.2   Simulation of a simple Miller OTA

In order to get an impression of the influence of the parasitic effects described above, a simple CMOS Miller operational amplifier has been implemented using the programmable transistor cells. Fig. 6 shows the implementation in the cell array (referred to as Cell-Op) as well as the equivalent circuit that results from

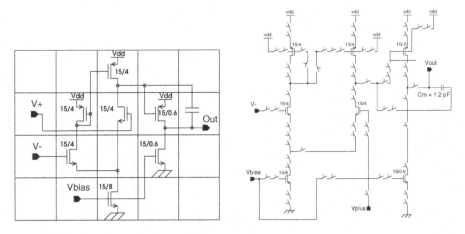

Fig. 6. Simulated operational amplifiers: **Left:** Implementation in the PTA chip. **Right:** Model of the circuit shown on the left including only the closed switches, here drawn as open ones for recognizability.

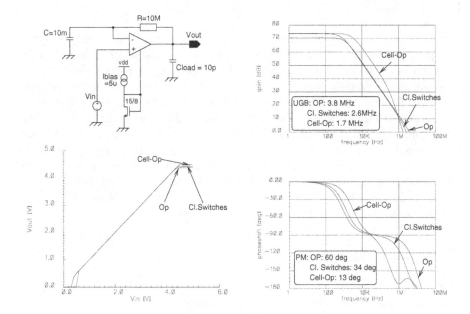

Fig. 7. Simulation results for the circuits shown in Fig. 6 and the original amplifier without switches: **Left: Top:** Test setup for the simulation. **Bottom:** DC-Responses **Right:**Results of the AC analysis: **Top:** Gain versus frequency (Bode plot). **Bottom:** Phase shift versus frequency.

disregarding all opened switches (referred to as Cl.Switches).

Both, ac and dc simulations[2], were carried out for three different implementations of the Miller OTA, namely the two circuits shown in Fig. 6 and the same Opamp without any switches (referred to as Op). The test setup for the simulation is shown in Fig. 7 together with the according dc and ac responses. For the dc analysis, where the Opamps are hooked up in a voltage follower configuration all three show the same desired behaviour for voltages between 0.5 V and 4.5 V (the region of interest).

The results of the ac analysis given in Fig. 7 show that the frequency response is degraded significantly by the introduction of both, the closed and the additional open switches (The according unity gain bandwidths (UGB) and phase margins (PM) are listed in the inset of the frequency plots.). This is due to the parasitic resistors and capacities introduced with the additional switches.

What can be learned from this? First of all, the maximum bandwith of the PTA chip is of the order of MHz. Secondly, since the GA will have to deal with the parasitic capacitances, the evolved solutions will require a different frequency compensation than their ideal counterparts.

---

[2] All simulations were made using the circuit simulator spectreS.

6 Summary

A new research tool for evolution of analog VLSI circuits has been presented. The proposed system featuring a programmable transistor array, which will be designed in a 0.6 μm CMOS process, will be especially suited to host hardware evolution with intrinsic fitness evaluation. Advantages are the fast fitness evaluation and therefore high throughput of individuals as well as the possibility to create a selection pressure towards robustness against variations of the environment and chip imperfections. Furthermore analyzability of the evolved circuits is enhanced by the implementation of local voltage, current and temperature sensors. First simulation results prove the feasibility of the chosen approach. The flexible design allows the test of different signal processing concepts with different chips easily derived from the chip presented here. Submission of the chip is planned for the first quarter of the year 2000.

7 Acknowledgement

This work is supported by the Ministerium für Wissenschaft, Forschung und Kunst, Baden-Württemberg, Stuttgart, Germany.

References

1. M. Loose, K. Meier, J. Schemmel: Self-calibrating logarithmic CMOS image sensor with single chip camera functionality, *IEEE Workshop on CCDs and Advanced Image Sensors, Karuizawa, 1999, R27*

2. Thompson, A.: An evolved circuit, intrinsic in silicon, entwined with physics. In Higuchi, T., & Iwata, M. (Eds.), *Proc. 1st Int. Conf. on Evolvable Systems (ICES'96)*, LNCS 1259, pp. 390-405. Springer-Verlag.

3. Thompson, A.: On the Automatic Design of Robust Electronics Through Artificial Evolution, In: Proc. 2nd Int. Conf. on Evolvable Systems: *From biology to hardware (ICES98)*, M. Sipper et al., Eds. , pp13-24, Springer-Verlag,1998.

4. R. Zebulum, M. Vellasco ,M. Pacheco,: Analog Circuits Evolution in Extrinsic and Intrinsic Modes, In: Proc. 2nd Int. Conf. on Evolvable Systems: *From biology to hardware (ICES98)*, M. Sipper et al., Eds., pp 154-165, Springer-Verlag,1998.

5. Layzell, P.: Reducing Hardware Evolution's Dependency on FPGAs, In *7th Int. Conf. on Microelectronics for Neural, Fuzzy and Bio-inspired Systems (MicroNeuro '99)*, IEEE Computer Society, CA. April 1999.

6. M. Murakawa, S. Yoshizawa, T. Adachi, S. Suzuki, K. Takasuka, M. Iwata, T. Higuchi: Analogue EHW Chip for Intermediate Frequency Filters, In: Proc. 2nd Int. Conf. on Evolvable Systems: *From biology to hardware (ICES98)*, M. Sipper et al., Eds., pp 134-143, Springer-Verlag,1998.

7. Stoica, A.: Toward Evolvable Hardware Chips: Experiments with a Programmable Transistor Array. In *7th Int. Conf. on Microelectronics for Neural, Fuzzy and Bio-inspired Systems (MicroNeuro '99)*, IEEE Computer Society, CA. April 1999.

8. K. R. Laker, W. M. C. Sansen: *Design of analog integrated circuits and systems*, pp 17-23, McGraw-Hill, Inc. 1994

Understanding Inherent Qualities of Evolved Circuits: Evolutionary History as a Predictor of Fault Tolerance

Paul Layzell and Adrian Thompson

Centre for Computational Neuroscience and Robotics, Centre for the Study of Evolution
School of Cognitive Sciences, University of Sussex, Brighton BN1 9QH, UK.
paulla, adrianth @cogs.susx.ac.uk

Abstract. Electronic circuits exhibit *inherent* qualities, which are due to the nature of the design process rather than any explicit behavioural specifications. Circuits designed by artificial evolution can exhibit very different inherent qualities to those designed by humans using conventional techniques. It is argued that some inherent qualities arising from the evolutionary approach can be beneficial if they are understood. As a case study, the paper seeks to determine the underlying mechanisms that produce one possible inherent quality, 'Populational Fault Tolerance', by using various strategies including the observation of constituent components used throughout evolutionary history. The strategies are applied to over 80 evolved circuits and provide strong evidence to support an hypothesis – that Population Fault Tolerance arises from the incremental nature of the evolutionary design process. The hypothesis is used to predict whether a given fault should manifest the quality, and is accurate in over 80% of cases.

1   Introduction

The well known techniques of abstraction and analysis in conventional circuit design are capable in principle of producing circuits which conform perfectly to a given set of behavioural specifications the very first time they are implemented in hardware or simulation. In contrast, circuit design by artificial evolution requires the implementation of *every* candidate design for evaluation, and proceeds by making gradual improvements to those candidates that most closely adhere to the desired specifications. When compared with the conventional method, the evolutionary approach may be deemed an *incremental* process. Even where in practice, conventional design incorporates an apparently similar phase of trial-and-error, any modifications are made with an *expectation*, based on knowledge or analysis, of their consequent effect on the circuit's behaviour. Hence, applicable modifications are limited to a subset containing those that can be analysed or understood. However, modifications by evolutionary operators are made with no *a priori* expectations of their consequences. Evolution does not need to analyse its modifications – it only needs to *observe* the subsequent circuit behaviour. Hence, its modifications come from a very different subset, containing all those permitted by the architecture, genotype/phenotype map-

J. Miller et al. (Eds.): ICES 2000, LNCS 1801, pp. 133–144, 2000.
© Springer-Verlag Berlin Heidelberg 2000

ping, and genetic operators, regardless of whether they can be analysed or under-stood. It is sometimes desirable to restrict this subset to include only modifications that adhere to conventional design rules, especially in digital design, for example [4,10]. However, where this is not the case, evolution is free to exploit components and architectures in new and unusual ways [15].

The fundamentally different nature of these two design processes is reflected in the kinds of circuits they produce. Both exhibit qualities that are *inherent,* that is, due to the nature of the design process, not to any explicit design specifications. For ex-ample, conventional circuits generally have a clear *functional decomposition,* whereas evolved ones may not [14]. Some inherent qualities may be undesirable, for example where evolution exploits parasitic properties, making it difficult to repro-duce a circuit on different hardware [8,15]. This paper attempts to isolate a poten-tially highly desirable quality inherent in circuits designed by Genetic Algorithms (GA). Hereafter referred to as 'Populational Fault Tolerance' (PFT), it is the potential for a population of evolved circuits to contain an individual which adequately per-forms a task in the presence of a fault that renders the previously best individual useless. If the fault is persistent, the new 'best' individual may be used to seed fur-ther evolution, possibly attaining performance equal to that before the fault occurred, in a fraction of the time it took to evolve the circuit from scratch.

The next section explains the value of inherent qualities of evolved circuits in more detail, suggests techniques for better understanding them, and describes the motivation for the current study of PFT. An experimental section then analyses over 80 test cases on which faults are inflicted, to determine whether PFT is truly an in-herent quality of evolved circuits, and examines various hypotheses for the underly-ing mechanisms that produce it. The results produced by the final experiment, which examines how constituent components were added to the design during evolutionary history, allows us to predict whether a given component will manifest PFT when faulty, with a high degree of confidence. The conclusion is that PFT is indeed an inherent and predictable quality of certain generic classes of evolved circuit.

2   Motivation

As the field of hardware evolution (HE) progresses as an engineering methodology, it becomes increasingly important to ascertain and understand areas at which it excels. Currently, a large proportion of the field is devoted to areas that are problematic for conventional design, for example [6,12], but comparatively few analytical tools are available to help determine whether such areas are any less problematic for HE. Iso-lating and understanding inherent qualities is a beneficial approach to making such determinations, since many inherent qualities are likely to be common to general classes of evolutionary algorithm rather than just resulting from specific nuances of some particular method. Furthermore, inherent qualities by definition arise 'for free' from the design process. Where an inherent quality is desirable, an understanding of the mechanisms that produce it may permit tailoring the design strategy to further encourage it. However, there are no universal methods for providing such understanding. Useful insights are gained by analysis of evolutionary runs performed on fitness landscapes such as Royal Road [16], NK [5], and NKp [2], for example [13,17]. All of these share certain characteristics with real design space, but otherwise bear little resemblance to it. An alternative first step is to analyse the evolved circuits, for which several useful tactics are suggested in [14]. One such tactic is to observe the behaviour and internal operation of a circuit's evolutionary ancestors, and will prove particularly powerful in the following study.

The phenomenon of PFT and the notion that it may be an inherent quality of evolved circuits was first postulated in [7]. A preliminary study was conducted [9] following observations of similar effects in previously evolved circuits and in other fields, for example [3]. It provided enough evidence to justify the more detailed study documented here. While it is hoped that PFT may eventually supplement existing fault tolerance techniques [1,11], the primary aim of this paper is to *understand* it, not to assess its usefulness.

3   The Experimental Framework

The study in the following sections comprises two elements. The first is an empirical study consisting of many evolutionary runs of several different tasks. On completing the runs, hardware faults are effected, and data taken concerning the subsequent performance of individuals in the final population, and some of their ancestors. The second element is an analysis of the evolved circuits to determine the relative importance of their constituent components, thus providing an indication of the severity of a given fault. The runs were carried out in hardware using an Evolvable Motherboard (EM) [7], a re-configurable platform on which plug-in components can be interconnected using programmable analogue switches, which are arranged in a triangular matrix upon a grid of wires (figure 1). The EM is ideal for such a study as faults can be induced by physically removing the plug-in components (and replacing them with faulty ones if necessary). Alternatively, the connections can be re-routed by software, which allows effective removal or substitution to be carried out automatically. Both methods were used in the following experiments. Only bipolar transistors (and the analogue switches in a static state) were provided as constituent components, up to five each of PNP and NPN. The tasks used in the study - digital inverters, inverting amplifiers, and oscillators - were chosen as test cases for the following reasons:

1) They are simple enough to evolve many times from scratch within a tractable time-scale
2) They pose increasingly difficult challenges for fault tolerance. Faults are expected to be particularly troublesome for oscillators if they rely for their operation on chains of components within a loop
3) They are expected to exploit the transistors in different ways: the inverters to use saturation mode, the amplifiers to use linear mode, and the oscillators to exploit parasitic capacitance.

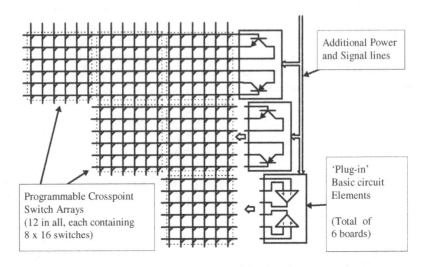

Figure 1. A simplified representation of the Evolvable Motherboard

The tasks were evolved using a rank-based, generational GA with elitism, and an initially random population of 50 individuals. Genetic operators were mutation (probability 0.01 per bit) and single-point crossover (probability 1.0). The genotype is mapped to the EM such that few constraints are imposed on the way components can be exploited or configured. Each motherboard row is assigned a corresponding column. The genotype is a linear bitstring representing the switches a row at a time. For each row, one bit specifies connection to the corresponding column, followed by the column number (6 bits) and connection bits for up to n additional switches. A more detailed explanation of this mapping can be found in [7]. The runs in this paper used either 48 EM rows/columns with n set to 3, or 40 rows/columns, n set to 4, giving genotype lengths of 1056 and 1160 bits respectively.

The digital inverter was evaluated using a series of 10 test inputs containing 5 '1s' and 5 '0s' (logical Highs and Lows, respectively) applied sequentially in alternating order. Fitness F_{INV} was scored according to equation 1:

$$F_{INV} = \frac{1}{5}\left\{\left(\sum_{t \in S_L} v_t\right) - \left(\sum_{t \in S_H} v_t\right)\right\} \qquad (1)$$

where t signifies the test input number, S_L and S_H the set of Low and High test inputs respectively, and v_t ($t=1,2,..10$) the measured output voltage corresponding to test number t.

The inverting amplifiers were evaluated by applying to the input a 1kHz sine wave of 2mV peak-to-peak amplitude, with a d.c. offset of half the supply voltage. Both input and output were monitored using an A/D converter, and 500 samples of each taken at 10us intervals. Fitness F_{AMP} was scored according to equation 2, where

a is the desired amplification factor, set to -500 for all runs, V_{in_i} and V_{out_i} are the ith input and output voltage measurements respectively, and O_{in} and O_{out} are the d.c. offset voltages of input and output respectively. Note that the amplifiers were not intended for practical use, since equation 2 does not take phase shift between input and output into account, and the single-frequency input does not allow frequency response to be evaluated.

$$F_{AMP} = -\frac{1}{500}\sum_{i=1}^{500}\left|a(V_{in_i} - O_{in}) - (V_{out_i} - O_{out})\right| \tag{2}$$

The oscillators were evaluated using analogue hardware to convert frequency and amplitude to d.c. voltages, which were then sampled using an A/D converter. Using this method, rather than sampling the circuit's output directly, ensures that oscillation frequencies higher than the A/D card's sample rate do not produce erroneous results due to aliasing. A set of 20 measurements each of amplitude and frequency was made over a sample period of 2ms for each candidate circuit. Fitness F_{OSC} was scored according to equation 3:

$$F_{OSC} = \bar{a} + \frac{f_{min}}{f_{max}}\left(2f_{target} - \left|f_{target} - \bar{f}\right|\right)$$

$$F_{OSC} = \bar{a} \quad \text{if } \bar{f} < 60Hz \tag{3}$$

$\bar{a}$ and $\bar{f}$ represent respectively output amplitude and frequency averaged over the 20 samples, f_{min} and f_{max} respectively the minimum and maximum frequencies measured during the sample period, and f_{target} the target frequency. The ratio of minimum and maximum frequencies serves to penalise individuals with varying output frequency. For equation 3 to be a smooth function, the frequency to voltage converter must be adjusted so that its output voltage corresponding to the target frequency is equal to half it's maximum output voltage. The evolved oscillators attained the target frequency (approximately 25kHz) to within 1%, with a minimum amplitude of 100mV.

For all tasks, the supply voltage was set to +2.8V, and the circuit output was buffered by an FET-input operational amplifier before entering the A/D converter.

3.1 Does PFT Actually Exist?

The first experiment was devised to give an idea of the extent (if any) to which PFT occurs for the test cases described above. Multiple runs were conducted until each task had been successfully evolved 20 times. At the end of each run, the population had reached a high degree of genetic convergence. Faults were effected by removing

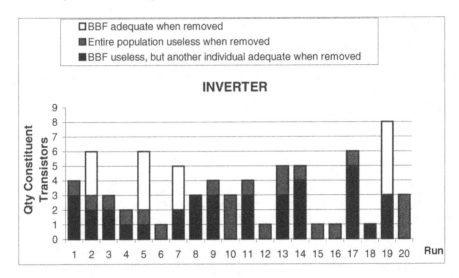

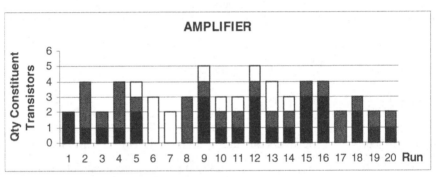

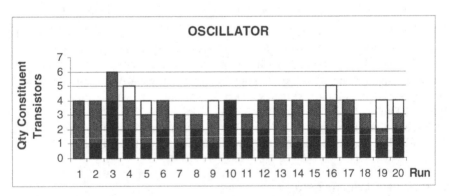

Figure 2. Graphs showing constituent transistors for each of 60 test runs. The bars are shaded according to the effect a transistor's removal has on the population's performance (see text).

transistors used by the best individual in the final population, one at a time. This individual is hereafter referred to as the BBF (Best Before Fault) individual. The removal of a transistor is equivalent to a fault common with bipolar transistors – both PN junctions becoming open-circuit. The constituent transistors were then classified by analysing their role in the evolved circuits. Transistors whose removal did not affect the BBF individual's fitness were not considered constituent, even if they seemed 'meaningfully' wired in the schematic. Transistors wired such that their collector-emitter current was clearly modulated by base voltage (in other words, the conventional wiring of a transistor) were termed *active*, and the rest *passive*. Faults in both types may be equally severe in terms of their effect on the performance of a single individual. However, if passive transistors are configured as 'wires', a population is likely to contain individuals of similar fitness that do not use these transistors. We wish to establish whether PFT is limited to such simple cases.

The extent to which reduced performance of a circuit in the presence of a fault can be considered *adequate* is clearly dependent on the task. Here, the minimum adequate performance is defined as corresponding to a fitness score significantly above that obtained by a trivial circuit giving constant output.

Such a circuit would be an appropriate seed for further evolution, even if it were no longer suitable for the intended task in its existing state.

Results are shown in the bar graphs of figure 2, which show the quantity of constituent transistors for each run. Constituent transistors are shaded according to the effect their removal had on the final population's performance. The white areas indicate the numbers of transistors whose removal did not seriously worsen the performance of the BBF individual. The grey areas indicate the numbers of transistors whose removal resulted in inadequate performance from the entire population. Finally, the black areas indicate the numbers of transistors whose removal resulted in inadequate performance from the BBF individual, but adequate performance from another member of the population. These latter are faults for which PFT occurs. According to the definition of PFT, we are only concerned with the black and grey areas: 'faults' that rendered the BBF individual useless. The larger the ratio of black versus grey for a given run, the better the PFT.

The graphs show enough evidence of PFT for the test cases to justify further analysis. However, there is one significant concern. In nearly all cases where the entire population was rendered useless, the transistor removed had been active rather than passive. At first sight, faults in active transistors may be too severe for PFT to reliably occur, in which case its potential as a strategy for producing robust populations of circuits is limited. Fortunately, the next sections show that this is unlikely to be the case.

3.2   Examining various hypotheses to explain how PFT occurs

Having established evidence for the existence of PFT, we now seek to understand the underlying mechanisms that produce it. The experiments described so far have been carried out using a particular evolvable architecture. We cannot predict the extent to which PFT might arise using other architectures that have more or less redundancy, or different constraints on component use. Therefore, we simplify the search to explain PFT by confining further analysis to apply only to the architecture on which the test cases were evolved. Given a constant architecture, we can propose several hypotheses to explain how PFT occurs. Those that appear most likely are listed below:

H1) Although the final GA population was converged, it still contained diverse solutions.

H2) PFT is the result of the incremental nature of the evolutionary design process, as described in the introduction. Configurations of ancestor circuits (which did not make use of the transistors whose removal resulted in PFT) still exist relatively undisturbed in the final genotypes.

H3) PFT is not an inherent quality of evolved circuits. It is a general property of any design implemented on this architecture. Hence, a population of mutants of conventional circuits could be expected to give similar results.

H4) PFT is the result of some peculiarity of the evolutionary algorithm employed, for example the crossover operator.

H5) The transistors whose removal manifested PFT were not used in a functional capacity, but merely acted as resistances. They could be bypassed by mutation with little difference in circuit performance.

A thorough comparison of the BBF individuals' schematics with those that performed better in the presence of a fault suggest that H5 is rarely the case. Indeed, the oscillator circuits consisted mainly of active transistors whose removal very often manifested PFT. The following experiment puts H1 and H4 to the test. In place of a GA, 10 runs each of inverter and amplifier circuits were designed using a hill-climbing algorithm, using identical mutation rates and mapping to those used by the GA. It was not possible to complete 10 oscillators using this method.

On completing the runs, 49 copies of the final individual were mutated (with GA mutation rates) to form a population of 50 including the final individual, and the previous experiment was repeated using these populations. The use of a hill-climber (essentially an elitist GA with a population of two, and no crossover operator) helps to determine any contribution to PFT of population diversity and crossover. Figure 3 shows the results. The inverter is not much help here, as very few faults resulted in the BBF individual (the un-mutated hill-climber) performing inadequately[1], but the amplifier shows similar results to the GA. Indeed for both tasks, a maximum of only one constituent transistor rendered the entire population useless. The implication is that any PFT observed here was due to neither crossover (H4), nor any more diversity than one set of mutants of a single individual (H1).

---

[1] The reasons for this are known. The hill-climber tended to produce inverter circuits with two transistors in parallel, the fractionally higher voltage swing attained giving slightly higher fitness. Hence removing either transistor made very little difference to performance.

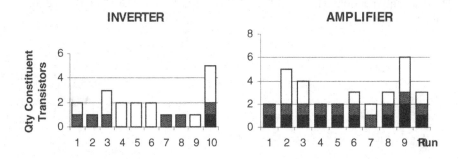

Figure 3. Constituent transistors of 20 hill-climber runs, shaded as for the graphs of figure 2.

3.3   Using Evolutionary History to Predict PFT

While hypotheses H1, H4 and H5 are not conclusively ruled out (particularly H4), the evidence against them makes the remaining two the most likely candidates. H3 proposes that PFT may not be due to the evolutionary design process at all. One approach to determine whether this is the case could be to describe a number of conventionally designed circuits by genotypes with identical phenotype mapping to that used for the first experiment. A population of mutants could then be made and instantiated in hardware, and the previous experiment repeated. However, there are a number of insuperable difficulties associated with this approach. Firstly, hand-designing enough different circuits using only transistors and motherboard switches to provide statistically valid data would be extremely difficult. Additionally, care would have to be taken to extend the redundancy which evolution was afforded to the conventionally designed equivalents (the evolved circuits used on average only about half of the available transistors). Instead, consider H2: If the transistors whose removal manifests PFT are indeed later additions to some early ancestral prototype whose configuration remains largely intact, then there should be a correlation between such transistors and the point in evolutionary history at which they were incorporated. Transistors adopted by the very earliest solutions would be less likely to manifest PFT upon removal than those adopted at the latter stages. The final experiment was devised to test for such a correlation.

Having completed an evolutionary run, the best individual of every generation is subjected to the removal of transistors one at a time, and re-evaluated. If the individual's fitness is consequently reduced, then the removed transistor can be assumed constituent for that generation. By making the test for every generation, it is possible to build up a profile of any particular transistor's importance to the best individual at all stages during the evolutionary history of that run. Since performing this experiment requires removing and replacing transistors many times, it was automated by re-routing the EM connections from the transistor in question to unused columns.

This was only possible with 10 of the 20 runs per task, since the other 10 used all 48 of the EM columns.

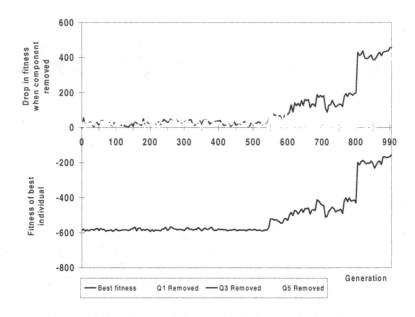

Figure 4. Plots indicating which stage of evolutionary history the constituent transistors were incorporated into the design of an evolving amplifier.

Figure 4 shows a typical result, in this case for an evolved amplifier. The lower trace is a plot of the best individual's fitness with all transistors intact. The upper traces are of the *reduction* in the best individual's fitness upon removal of a given transistor. Hence a non-constituent transistor will not affect the best individual's fitness when removed, giving a reduction of zero. Likewise, a constituent one will reduce the best individual's fitness when removed, giving a positive value. For clarity, plots for only three transistors are displayed. The graphs clearly show how the respective transistors (labelled 'Q') were incorporated into the design as evolution progressed. The first prototype solution (generation 550) required Qs 3 and 5, however Q5 (possibly used as a 'wire') was quickly dropped while Q3 remained essential for the prototype's operation. Q1 was incorporated into the design at generation 730 and became an essential component from generation 800 onwards, as did Q5 from generation 950.

Disregarding measurement noise, the trace for Q3 is virtually identical to that for best fitness throughout the evolutionary history. The same effect was observed in *every one* of the 30 runs: One or two of the transistors forming the prototype remain an essential part of the design while it is improved by the gradual addition of other transistors. We are now in a position to test hypothesis H2. If H2 is correct, then we can predict that transistors whose fitness reduction traces are identical to the plot for best fitness will *not* manifest PFT when removed, while any others whose removal renders the BBF individual useless, will. Astonishingly, this prediction turns out to be correct for 26 of 30 cases (inverter), 27 of 29 cases (amplifier), and 28 of 35 cases (oscillator), or 87%, 93%, and 80% correct, respectively. The correlation is too close to be chance: While the other hypotheses may still have some small role, H3 surely cannot be true: PFT is established as an inherent quality.

4   Conclusion

The evolutionary history of candidate circuit designs showed clearly that once the GA had found a prototype circuit giving significantly better fitness than constant output, all subsequent designs were based on it. The accuracy with which it was possible to make predictions of PFT based on which transistors were essential components of the prototype and which were later additions, is very strong evidence in favour of earlier circuits' configurations remaining largely intact. If this is the case, then PFT is truly an inherent quality of evolved circuits produced by the evolutionary algorithms and evolvable architecture described in this paper. Indeed, it is likely that PFT is inherent in a broader generic class of evolved circuits. Using the knowledge of PFT gained by this study, we can speculate that such a class might contain circuits designed by many incremental strategies, which gradually incorporate additional components into a basic initial prototype. We can also speculate that PFT would be less likely to occur for on-line evolution in a varying environment: Although later designs may be gradual improvements of a largely unchanged ancestor, the ancestor itself would be of little use were it a solution to a different problem. There is room for further understanding and improving PFT. Future work includes investigation of PFT for different evolvable architectures, and experiments to determine any relationship between the qualitative drop in performance caused by a faulty component and the time in evolutionary history at which that component became constituent.

PFT is one of possibly many desirable inherent qualities yet to be yielded by the evolutionary design process. Isolating and understanding others can only increase hardware evolution's effectiveness as an engineering tool.

Acknowledgements

This research was funded by British Telecommunications PLC. The authors would like to thank the anonymous referees for their comments and suggestions.

References

1. Avizienis, A.: Toward Systematic Design of Fault-Tolerant Systems, IEEE Computer, Comp. Soc. Press, (April 1997) pp.51-58
2. Barnett, L.: Riggedness and Neutrality – The NKp Family of Fitness Landscapes, *Proc, 6th Int. Conference on Artificial Life.* MIT Press. (1998) pp.18-27
3. Dittrich, P.,Burgel, A., Banzhaf, W.: Learning to Move a Robot with Random Morphology. In Husbands & Meyer (Eds.) *Proc. 1st Eur. Workshop, EvoRobot98,* Vol. 1478 of LNCS. Springer-Verlag. 1998. pp 165-178.
4. Kajitani, I., Hoshino, T., Nishikawa, D. et al.: A Gate-Level EHW Chip: Implementing GA Operations and Reconfigurable Hardware on a Single LSI. In Sipper, M., Mange, D., Perez-Uribe, A. (Eds.). *Proc. 2nd Int. Conf. on Evolvable Systems: From Biology to Hardware,* Vol. 1478 of LNCS. Springer-Verlag. (1998). pp.1-12.
5. Kauffman, S.: *The Origins of Order.* Oxford University Press. (1993).
6. Higuchi, T. and Kajihara, N.: Evolvable Hardware Chips for Industrial Applications. In Yao, X. (Ed.).*Communications of the ACM.* Vol. 42, no. 4. (April 1999). pp. 60-66.
7. Layzell, P.: A New Research Tool for Intrinsic Hardware Evolution. In Sipper, M., Mange, D., Perez-Uribe, A. (Eds.). *Proc. 2nd Int. Conf. on Evolvable Systems: From Biology to Hardware,* Vol.1478 of LNCS. Springer-Verlag. (1998). pp.47-56.
8. Layzell, P.: Reducing Hardware Evolution's Dependency on FPGAs. *Proc. 7th Int. Conf. Microelectronics for Neural, Fuzzy, and Bio-Inspired Systems.* IEEE Computer Society, CA. (April 1999). pp171-8.
9. Layzell, P.: Inherent Qualities of Circuits Designed by Artificial Evolution: A Preliminary Study of Populational Fault Tolerance. In Stoica, A., Keymeulen, D., Lohn, J. (Eds.) *Proc. 1st NASA/DOD Workshop on Evolvable Hardware.* IEEE Comp. Soc. Press (1999) pp.85-86.
10. Miller, J. and Thomson, P.: Aspects of Digital Evolution: Geometry and Learning. In Sipper, M., Mange, D., Perez-Uribe, A. (Eds.). *Proc. 2nd Int. Conf. on Evolvable Systems: From Biology to Hardware,* Vol.1478 of LNCS. Springer-Verlag. (1998). pp.25-35.
11. Ortega, C., Tyrell, A.: Reliability Analysis in Self-Repairing Embryonic Systems. *Proc. 1st NASA/DOD Workshop on Evolvable Hardware.* IEEE Comp. Soc. Press (1999) pp.85-86.
12. Sipper, M., Mange, D., Perez-Uribe, A. (Eds.): *Proc. 2nd Int. Conf. on Evolvable Systems: From Biology to Hardware,* Vol.1478 of LNCS. Springer-Verlag. (1998).
13. Thompson, A.: Evolving fault tolerant systems. In *Proc. 1st IEE/IEEEE Int. Conf. On Genetic Algorithms in Engineering Systems: Innovations and Applications (GALESIA '95),.* IEE Conf. Publication No. 414. (1995a) 524-529
14. Thompson, A. & Layzell, P.: Analysis of Unconventional Evolved Electronics. In Yao, X. (Ed.).*Communications of the ACM.* Vol. 42, no. 4. (April 1999). pp. 71-79.
15. Thompson, A., Layzell, P., Zebulum, R.: Explorations in Design Space: Unconventional Electronics Design Through Artificial Evolution. *IEEE Trans. Evolutionary Computation,* vol. 3, no. 3. (Sept. 1999) pp.167-196
16. Van Nimwegen, E., Crutchfield, J., Mitchell, M.: Statistical Dynamics of the Royal Road Genetic Algorithm. In Eiben, E. Rudolph, G. (Eds) *Special Issue on Evolutionary Computation, Theoretical Computer Science,* 229. (1999), pp.41-102.
17. Vassilev, V.,Miller, J., Fogarty, T.: Digital Circuit Evolution and Fitness Landscapes. In *Proc. Congress on Evolutionary Computation,* Piscataway, NJ. IEEE Press. (1999)

Comparison between Three Heuristic Algorithms to Repair a Large-Scale MIMD Computer

Philippe Millet[1,2], Jean-Claude Heudin[3]

[1] Université Paris-Sud Orsay, Institut d'Electronique Fondamentale, France
[2] Thomson-CSF AIRSYS, service RD/RTNC,
7-9 rue des Mathurins, 92221 Bagneux CEDEX, France
Philippe.Millet@AIRSYS.thomson.fr
[3] Pôle Universitaire Léonard de Vinci, 12, Institut International du Multimédia, 92916
Paris La Défense CEDEX, France
Jean-Claude.Heudin@devinci.fr

Abstract. This study tries to find a new task mapping using available resources of a partially broken MIMD machine based on an isochronous crossbar network. In this framework, we compare a genetic algorithm, a genetic programming algorithm and a simulated annealing algorithm. We show evidences that both genetic algorithms seem to be excellent to solve the problem.

1 Introduction

For few years, radar signal is processed numerically for better performances and flexibility [14]. New radar algorithms are efficient but need a lot of computer resources [9]. To computes such a signal real time, Thomson-CSF AIRSYS designed a parallel machine based on Digital Signal Processor (DSP) [15]. Such computers are very powerful but also very complex in terms of components and connections. This complexity lead to a large number of possible faults. At initialisation stage or at runtime, the machine can detect some of these faults and ask for someone to come and repair it. Since such huge radar systems can stand for months in inaccessible places, the system can remains out of order during a long period of time.

The aim of our work is to allow dynamic reconfigurations of the task-processor mapping when a fault is detected in the machine. Thus the tasks are mapped on using available resources, and unavailable resources are no longer used. Then the network is reconfigured, and the machine can restart.

J. Miller et al. (Eds.): ICES 2000, LNCS 1801, pp. 145–154, 2000.
© Springer-Verlag Berlin Heidelberg 2000

2   Problem description

2.1   Model

This study tries to find an algorithm to repair a Multiple Instruction Multiple Data (MIMD) machine. The architecture of such systems is often dedicated to the process they have to deal with. However, basically, a parallel computer is composed of computational elements that process the information (processors) and network links that spread the information [14][15]. Both elements can be affected by faults, but here we will focus on the processors. We will not describe the faults nor the method used to detect the faults, we assume that faulty processors are know by the algorithm that will try to repair. To test different of algorithms, we use a model of a MIMD machine based on a matrix, where each cell matches a processor.

00	01	02	03	04	-1
10	11	12	13	14	-1
20	21	22	23	24	-1
30	31	32	33	34	-1

Fig. 1. This is an array of processors. Each cell is a processor, each value is a task identifier. An identifier value of −1 means no task is mapped on this processor.

The values assigned to the cells are the task's identifications. A free cell has a task identification of -1. A broken processor means an unusable cell. As in the Travelling Salesman Problem (TSP) where one city should be visited only once [5], a processor can only be assigned a task, and one task can be assigned to only one processor. We will call this the rule of the TSP.

Fault tolerance problems are often solved using binary decision trees [7][8]. But in this study we are more interested in generating a new mapping of the tasks on the processors in response to a fault. By "killing" a processor, a fault reduces the number of usable resources. To replace the dead processors, the machine is added resource (processors, communication links...) only used for repair purpose.

Due to the complexity of the problem, we exclude "systematic algorithms" [17] and we look for heuristic algorithms. In this study, we will compare three different algorithms to solve the problem: (1) a Genetic Algorithm, (2) a Genetic Programming Algorithm, (3) a Simulated Annealing Algorithm.

We will analyse the speed and the quality of the mapping done by each algorithm. To determine the quality of a result, we will find the number of physical links used. The more links used, the less the quality.

2.2 Constraints and fitness evaluation

All candidate algorithms must match two constraints: (1) repair the machine by reallocating tasks from dead processors to living ones, (2) find a good mapping for the whole machine. We first try to solve the following problem: given an existing mapping, we tell the algorithm that a processor (or more) is dead. Thus, it has to find another mapping.

The higher the fitness, the closer to a solution that brings to a good mapping without using dead processors. The fitness function is very important because it manage the behaviour of the algorithm. Most of our work was the design of a function that can match all requirements of our problem. The number of links used by two tasks to spread data is the number of processors to go through from the transmitter to the receiver when using the shortest available path. Since maximum number of links the machine can use is not infinite, typical bad mapping will use too many links thus making it impossible to implement. The algorithms have to find a mapping that reduces the total number of links, and that doesn't use broken processors. In this study, we will assume each task sends data to its four originally connected neighbour. Thus, task '00' sends data to tasks '01' and '10'; and task '22' sends data to tasks '12', '23', '32' and '21'; and so on… A good mapping of the tasks is shown in problem description (cf. figure 1). In this matrix, the number of links used by two tasks to communicate is always 1. As an example, if tasks '01' and '14' are swapped, the number of needed links by task '00' to spread data to task '01' is equal to 5. Fitness function is based on this number of needed links.

$$Fitness = \frac{Max\_links - \sum_{each\_task}(Number\_of\_links\_needed)}{Max\_links - Min\_links} \times \frac{1}{Nb\_bad\_used} \tag{1}$$

Where Max_links is the total number of needed links when all communication distances are maximised, Min_links is the total number of links needed when all communication distances are minimised, and Nb_bad_used is the number of broken processors used. Considering the matrix shown in figure 1, the maximum distance between two tasks is 10 (one task at the top right corner and the other at the bottom left corner), the minimum distance is 1 (two tasks next one from the other). In this case, Max_links is equal to 1240, and Min_links is equal to 62.

3 Genetic Algorithm

3.1 Algorithm overview

This genetic algorithm uses data structured genes instead of binary genes [3]. The basic objects of our study are tasks and processors. In a previous study we chose to have an array of tasks and to assign a processor to each task [1]. However this not appear to be a natural way of coding such a problem, and we had to control many rules and parameters (such as the rule of the TSP) to make the algorithm really match the problem. In this study, we choose to have an array of processors as a chromosome, and to assign a task to each processor. Since we modelize the MIMD machine as a 2D matrix, we have to deal with 2D chromosomes. An example of such a chromosome is displayed by figure 1. The selection procedure used in this algorithm is the one described in a previous article [1]. This procedure uses a remainder stochastic sampling without replacement suggests by Booker [13].

3.2 Operators

The algorithm uses a 2D-crossover function [4]. In such a crossover, we select rectangular zones in the chromosome. The number of zones is given randomly. The size and position of the zones are also set randomly. We exchange the zones one by one [4]. When crossing, we have to pay attention to the rule of the TSP. So when we cross two tasks between two offspring, we exchange in each offspring the tasks to always keep the correct number of tasks.

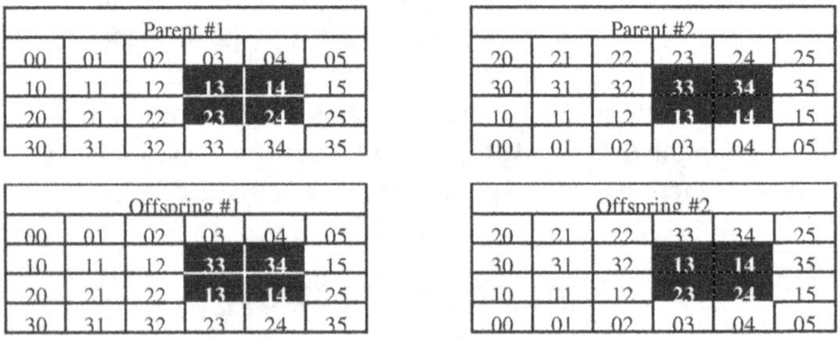

Parent #1

00	01	02	03	04	05
10	11	12	**13**	**14**	15
20	21	22	**23**	24	25
30	31	32	33	34	35

Parent #2

20	21	22	23	24	25
30	31	32	**33**	**34**	35
10	11	12	**13**	**14**	15
00	01	02	03	04	05

Offspring #1

00	01	02	03	04	05
10	11	12	**33**	**34**	15
20	21	22	**13**	**14**	25
30	31	32	23	24	35

Offspring #2

20	21	22	33	34	25
30	31	32	**13**	**14**	35
10	11	12	**23**	**24**	15
00	01	02	03	04	05

Fig. 2. The offspring resulting of a crossover of the two parents described above.

Portions of the genes, which are copied in crossover process, are modified with a mutation. We have 4 different mutation operators:

Simple copy error	Randomly, we modify the copied value by another one making a simple error in the copy process.
Shift	A Gene is selected randomly, and we shift the row or the column from this gene to the end of the chromosome.
Shift with Attraction	The same but the gene is selected in a biased wheel. The closer is a task to a dead processor the more it is present in the wheel.
Exchange	Two randomly selected genes (cells) are exchanged.

3.3 Parameters Summary

The values below have been set with commonly used values and then adjusted after few experimenting.

Parameter	Setting
Population size	200
Selection	Remainder stochastic sampling without replacement
Selection size	400
Max crossover zones	4
Crossover-frequency	0.6
Mutation-frequency (Simple copy error)	0.03
Mutation-frequency (Shift)	0.015
Mutation-frequency (Shift with Attraction)	0.003
Mutation-frequency (Exchange)	0.01
Termination-criterion	Exit after 10^6 generations

4   Genetic Programming Algorithm

4.1 Algorithm overview

The Genetic Programming algorithm evolves a program that modifies the mapping of the machine in order to repair it. A chromosome is made of a list of functions and arguments:

pointer to next cell
pointer to function to run
arguments of this function

Fig. 3. Cells of the evolving program.

The function called in each cell modifies the mapping. Each function is selected in a bag of available functions. The last function of the list does not modify, but evaluates the mapping done by the program described in the chromosome. The number of arguments depends on the function called. The size of the chromosome is not defined, chromosomes may have different sizes. We limited this size to ten times the expected size [3]. We randomly generate the first individuals. This include the length of each chromosome (maximum depth is 6), the functions called, the arguments for each function. The selection method used is tournament [3]. In this method, the selection takes place in a subset of the population. In this subset, the best individuals are allowed to reproduce with mutation. The offspring replace the worst individuals. In our program, the subset contains seven individuals, and the algorithm replaces the two worst individuals.

4.2 Operators

When two individuals are selected, they may cross their genes. This crossover randomly cuts each selected chromosome and exchanges the functions from those points [6]. Crossing an individual with itself is not forbidden [3]. The implemented mutations will act on different part of the evolving program:

Function Mutation	Will replace a function in the program with another one selected randomly in the bag of available functions.
Tree Mutation	Will exchange two parts of the chromosome.
Arguments Mutation	Will modify randomly the argument of a function in the program.

The basic set of functions used by the algorithms implements the elementary used to modify the mapping. Such modifiers look like the mutation functions created in the genetic algorithm. We implemented 5 functions:

Exchange	Exchanges two tasks A and B in the machine. Argument 1: position of a task A in the machine. Argument 2: position of a task B in the machine.
Shift columns	Shifts the tasks on the same row from task A. Argument: position of a task A in the machine.
Shift rows	Shifts the tasks on the same column from task A. Argument: position of a task A in the machine.
Roll columns	Rolls the tasks on the same row from task A. Argument: position of a task A in the machine.
Roll rows	Rolls the tasks on the same column from task A. Argument: position of a task A in the machine.

4.3 Parameters Summary

Parameter	Setting
Population size	100
Selection	Tournament
Tournament size	7, kills 2
Crossover-frequency	1
Mutation-frequency (tree)	0.1
Mutation-frequency (functions)	0.1
Mutation-frequency (argument)	0.09
Maximum program size	50
Maximum initial size	6
Instruction set	Exchange, Shift column, Roll column, Shift row, Roll row.
Termination-criterion	Exit after 25000 Generations

5   Simulated Annealing

5.1   Algorithm overview

Simulated annealing has been used many times for placement problems [11][16]. Since each task has two co-ordinated (column and row), to find a mapping for N tasks, this algorithm has to find the best values for 2N parameters. This way the simulated annealing works on a 2D matrix without really using it. The annealing schedule for the temperature, T, from a starting temperature T0, a number of parameters D, is [2]:

$$T = T0 \exp(-10^{-5}\wedge(1/D)). \tag{2}$$

Our cost function can evaluate the mapping of a task compared to the mapping of the others. But only one task cannot know if it is well mapped in the machine. This lead the algorithm to completely change the mapping, and the final result is not acceptable. In our problem, some tasks has to be mapped on a specific processor (because of the specific capacity of this processor). Thus we forced the mapping of some tasks (say the first column). The algorithm can choose any value to map the other tasks. We use a Gaussian distribution to travel through the search space. The lower the temperature, the sharper the Gaussian becomes around the previous position. To find the next position, we randomly select it according to the probability given by the Gaussian.

During the process, we keep in memory the best mapping found. This mapping is the final result shown when the algorithm stops [10].

5.2   Parameters Summary

Parameter	Setting
Number of parameters	200
Temperature schedule	T0 exp(-10^{-5}^(1/D)).
Distribution used	Gaussian
Starting parameters	Each parameter is set following the specified
Termination-criterion	Temperature too low to modify the mapping.

6 Results

6.1 Repair a broken machine

The four arrays displayed by figure 4 show the result of each algorithm trying to solve the first problem. The dead processor is the black one. We chose to map 90 tasks on 100 processors because it is about the size of typical radar applications [1].

Each algorithm solved this problem. The genetic programming algorithm and the genetic algorithm gave about the same result, while the simulated annealing gave the worst result, but a good result if we look at it only. When solving the result they all save the global mapping.

Starting with mapping

00	01	02	03	04	05	06	07	08	-1
10	11	12	13	14	15	16	17	18	-1
20	21	22	23	24	25	26	27	28	-1
30	31	32	33	34	35	36	37	38	-1
40	41	42	43	**44**	45	46	47	48	-1
50	51	52	53	54	55	56	57	58	-1
60	61	62	63	64	65	66	67	68	-1
70	71	72	73	74	75	76	77	78	-1
80	81	82	83	84	85	86	87	88	-1
90	91	92	93	94	95	96	97	98	-1

Simulated Annealing

00	01	02	03	04	05	06	-1	08	07
10	11	22	13	12	16	15	18	-1	74
20	21	23	33	24	14	27	57	28	48
30	31	32	43	34	35	26	37	38	-1
40	61	62	53	**-1**	47	56	17	46	-1
50	71	72	44	54	55	36	25	45	-1
60	52	41	83	76	67	-1	66	68	58
70	51	42	65	64	75	73	77	87	-1
80	81	82	84	63	85	86	-1	88	97
90	91	92	93	94	95	96	98	78	-1

Genetic Programming

00	01	02	03	04	05	06	07	08	-1
10	11	12	13	14	15	16	17	18	-1
20	21	22	23	24	25	26	27	28	-1
30	31	32	33	34	35	36	37	38	-1
40	41	42	43	**-1**	45	46	47	48	44
50	51	52	53	54	55	56	57	58	-1
60	61	62	63	64	65	66	67	68	-1
70	71	72	73	74	75	76	77	78	-1
80	81	82	83	84	85	86	87	88	-1
90	91	92	93	94	95	96	97	98	-1

Genetic Algorithm

00	-1	01	02	03	04	05	06	07	08
10	-1	11	12	13	14	15	16	17	18
20	-1	21	22	23	24	25	26	27	28
30	-1	31	32	33	34	35	36	37	38
40	41	42	43	**-1**	44	45	46	47	48
50	-1	51	52	53	54	55	56	57	58
60	-1	61	62	63	64	65	66	67	68
70	-1	71	72	73	74	75	76	77	78
80	-1	81	82	83	84	85	86	87	88
90	-1	91	92	93	94	95	96	97	98

Fig. 4. Result of the three algorithms starting with the same mapping.

6.2 Quantitative results

In the next array, we compare the three algorithms.

	Genetic Algorithm	Genetic Programming	Simulated Annealing
Add a mapping modifier	Add a mutation, easy.	Add a function, easy	Impossible
Repairing a machine	Very good	Very good	Good
Time needed On an AMD K6-233 based PC	About 2 minutes	About 2 minutes	About 10 minutes
Fitness or cost function	Computes the number of link used, and gives a penalty when a task is mapped on a dead processor.		

The diagrams displayed by figure 5 and 6, show that simulated annealing is slower than the two others to find a solution, it requires ten times the time required by the genetic algorithm or the genetic programming algorithm. Both give very close results.

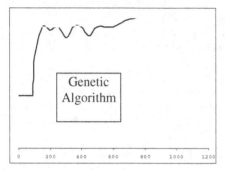
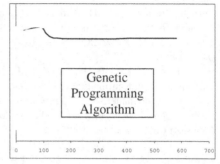

Fig. 5. We plotted in the same diagram the maximum fitness, the minimum fitness and the average fitness at each generation. The left diagram is done with the genetic algorithm while the other one is done with the genetic programming algorithm.

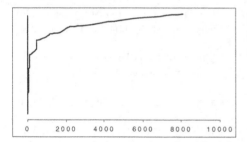

Fig. 6. This diagram shows the cost function value during simulated annealing algorithm iterations.

7 Conclusion

This work helps us to find an optimisation algorithm. Both the genetic algorithm and the genetic programming algorithm seem to be very good to solve the broken machine problem. In a next study, we will experiment with a randomly generated mapping. The algorithms will have to find a good mapping while starting from scratch. Then, the next step will be to implement the best algorithm within a real MIMD machine.

8 Acknowledgements

The authors would like to acknowledge Michel Coatrieux and Francois Giersch from Radar Traitement Numérique Department at Thomson-CSF AIRSYS. This research is supported by Thomson-CSF AIRSYS.

References

1. Millet, Ph., Heudin, J.-C.: Fault Tolerance of a Large-Scale MIMD Architecture Using a Genetic Algorithm. In: Sipper, M., Mange, D., Pérez-Uribe, Andrés. (eds.): Evolvable Systems: From Biology to Hardware. Lecture Notes in Computer Science, Vol. 1478. Springer-Verlag, Berlin Heidelberg New York (1998) 356–364
2. Adaptive Simulated Annealing (ASA) ©. http://www.ingber.com/ASA-README.html. . Lester Ingber Research, PO Box 06440, Wacker Dr PO Sears Tower, Chicago, IL 60606-0440 (1999)
3. Banzhaf, W., Nordin, P., Keller, E.R., Francone, F.D.: Genetic Programming An Introduction. Morgan Kaufmann Publisher, Inc., San Francisco, California. (1998) 97, 132-133
4. Kane, C., Schoenauer, M.: Optimisation topologique de formes par algorithmes génétiques. Tech. Rep. Centre de Mathématiques Appliquées, Ecole Polytechnique, 91128 Palaiseau Cedex, France. (1996)
5. Goldberg, D.E.: Algorithmes Génétiques Exploration, Optimisation et Apprentissage Automatique. Addison Wesley. (1991) 13-14, 186-194
6. Reiser, P.G.K.: Evolutionary Computation and the Tinkerer's Evolving Toolbox. In: Banzhaf, W., Poli, R., Schoenauer, M., Fogarty, T.C. (eds.): Genetic Programming. Lecture Notes in Computer Science, Vol. 1391. Springer-Verlag, Berlin Heidelberg New York (1998)
7. Andrews, J.D., Henry, J.J.: A Computerized Fault Tree Construction Methodology. In: Proceedings of the Institution of Mechanical Engineers. Part E, Journal of process mechanical engineering, Vol. 211. (1997) 171–183
8. Rauzy, A., Limnios, N.: A brief introduction to Binary Decision Diagrams. Binary decision Diagrams and Reliability. In: Journal européen des systèmes automatisés, Vol. 30. (1996) 1033–1050
9. Galati, G. (eds): Advanced RADAR Techniques and Systems. . Iee Radar, Sonar, Navigation and Avionics, No 4. Peter Peregrinus Ltd (1994)
10. Heckenroth, H., Siarry, P.: Optimisation par la méthode du recuit simulé du dessin d'un modèle conceptuel des données. In: AFCET/INTERFACES, No. 98. (1990) 11–23
11. Bonomi, E., Lutton, J.-L.: Le recuit simulé. In: POUR LA SCIENCE, No. 129. (1988) 68–77
12. Bonomi, E., Lutton, J.-L.: The N-City Travelling Salesman Problem: Statistical Mechanics and the Metropolis Algorithm. In: SIAM Review, Society for Industrial and Applied Mathematics, Vol. 26, No. 4. (1984)
13. Booker, L.B.: Intelligent Behaviour as an Adaptation to the Task Environment. Tech. Rep. No. 243. University of Michigan, Ann Arbor (1982)
14. OCTANE User manual. Technical Report n. 46 110 720 – 108, Thomson-CSF AIRSYS, Bagneux, France. (1997)
15. Specification B1, CAMARO Produit 96, Technical Report n. 46 110 432 - 306, Thomson-CSF AIRSYS, Bagneux, France. (1997)
16. Kirkpatrick, S., Gelatt, C.D., Jr., Vecchi, M.P.: Optimization by Simulated Annealing. In: SCIENCE, Vol. 220, No. 4598. (1983) 671–680
17. Schnecke, V., Vornberger, O.: Genetic Design of VLSI-Layouts. IEE Conf. Publication No. 414, Sheffield, U.K. (1995) 430-435

A Hardware Implementation of an Embryonic Architecture Using Virtex® FPGAs

Cesar Ortega and Andy Tyrrell

Department of Electronics
University of York
York, YO10 5DD, UK
{cesar, amt}@ohm.york.ac.uk

Abstract. This paper presents a new version of the MUXTREE embryonic cell suitable for implementation in a commercial Virtex® FPGA from Xilinx™. The main characteristic of the new cell is the structure of its memory. It is demonstrated that by implementing the memory as a look-up table, it is possible to synthesise an array of 25 cells in one XCV300 device. A frequency divider is presented as example of the application of embryonic arrays. After simulation, the circuit was downloaded to a Virtex FPGA. Results show that not only it is possible to implement many embryonic cells on one device, but also the reconfiguration strategies allow a level of fault tolerance to be achieved.

1 Introduction

The embryonics project introduces a new family of fault-tolerant field programmable gate arrays (FPGAs) inspired by mechanisms that take place during the embryonic development of multicellular organisms. Embryonics= Embryology + Electronics. Since its introduction, the embryonics project has evolved into different lines of research. [1, 2, 3]. At the University of York the fault tolerance characteristic of embryonic arrays has been investigated for some years now [4, 5, 6].

Embryonics was originally proposed as a new way of designing fault-tolerant FPGAs [7]. Hence, the ultimate goal of the embryonics researcher is to see his/her design integrated in silicon and early versions of the embryonics architecture have been tested using conventional FPGAs [1]. However, individual embryonic cells require large amounts of memory to store their configuration bits (genome). In a 16×16 array, approximately 4300 bits of memory **per cell** are required. Therefore, it becomes difficult to allocate more than one cell in a commercial programmable device. This paper presents a new version of the MUXTREE embryonic architecture that allows the implementation of complete arrays in a commercial FPGA.

Section 2 presents a brief introduction to the embryonics project and the evolution of the embryonics architecture studied at the University of York. A new approach to the design of the memory block is presented in section 3. Section 4 presents the design of a frequency divider in a 5×5 embryonic array and its implementation in a Virtex FPGA. Conclusions and proposals for future work are given in section 5.

J. Miller et al. (Eds.): ICES 2000, LNCS 1801, pp. 155–164, 2000.

2   The embryonics project

When biological multicellular organisms reproduce, a new individual is formed out of a single cell (the fertilised egg). During the days that follow the time of conception, the egg divides itself by a mechanism called mitosis. Through mitosis two cells with identical genetic material (DNA) are created. The new cells also divide, passing to every offspring a copy of the DNA that corresponds to the individual under development. At some point during their reproduction, cells differentiate into the different tissues that give shape to a complete healthy individual. Differentiation takes place according to "instructions" stored in the DNA (the genome). Different parts of the DNA are interpreted depending on the position of the cell within the embryo. Before differentiation cells are (to a certain extent) able to take over any function within the body because each one possess a complete copy of the genome [8].

The embryonics project transports these biological mechanisms to the world of electronic, programmable arrays. Figure 1 shows the generic architecture of an embryonic array. A detailed description of the cell can be found in [9]

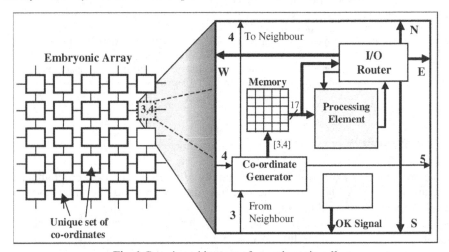

Fig. 1 Generic architecture of an embryonic cell

The configuration register of a cell defines the functionality of its processing element and routing resources. Every cell in an embryonic array stores not only its own configuration register, but also those of its neighbours. When differentiation takes place, cells select a configuration register according to their position within the array. Position is determined by a set of co-ordinates that are calculated from the co-ordinates of the nearest neighbours. Every embryonic cell performs self-checking by means of built-in self-test (BIST) logic. A detailed description of the self-testing mechanisms can be found in [7].

When a failure is detected, the faulty cell issues a status signal that propagates to the nearest neighbours. In response, some cells become transparent to the calculation of co-ordinates and consequently, they are logically eliminated from the array. Cells are eliminated according to the reconfiguration mechanism in use, *e.g.* cell elimination, row-elimination. The remaining cells recalculate their co-ordinates and select a new configuration register from their memory (genome). By selecting a new configuration register every cell performs a new function. Provided the amount of spare cells is sufficient to replace all the failing cells, the overall function of the original array is preserved [9, 1].

2.1   The MUXTREE embryonic cell

The MUXTREE is a variant of the generic embryonic cell where the processing element is a multiplexer, (henceforth referred to as the main multiplexer). The architecture of the generic MUXTREE cell requires a memory with capacity to store the configuration registers of all the cells in the array. Although this approach maps directly the concept of genome into the embryonic architecture, it is highly inefficient in the use of resources. In practice, a cell can only replace neighbouring faulty cells because of its limited connectivity. Therefore, there is little point in storing in a cell the configuration registers of cells that it will never replace. A detailed description of the generic MUXTREE architecture can be found in [9].

The fault tolerance characteristic of MUXTREE arrays has been the object of various studies at the Department of Electronics, University of York [4, 5, 6]. Considering that simplicity (small size) implies better reliability, the problem of memory size has been addressed following various strategies. To reduce the size of the memory inside each MUXTREE cell, a chromosomic approach was proposed in [5]. In an array performing row-elimination, faulty cells are substituted exclusively by their north neighbours. Hence, the chromosomic approach requires each cell to store only the configuration registers of all the cells in the corresponding column. Consequently, every cell in an array of size $n{\times}n$ has to store only $(n{+}1)$ configuration registers, instead of the $(n^2{+}1)$ required in the original proposal. The extra configuration register sets the transparent state that cells enter when they are eliminated.

3   A new design of the MUXTREE's memory element

As mentioned, the set of all the configuration registers in an embryonic array is called the genome. Once the genome of a particular application has been defined, it does not change throughout the useful life of the array. This implies that once defined, the memory behaves as a ROM or Look-Up Table (LUT); i.e. every set of co-ordinates (address) is associated to only one configuration word (data). This characteristic is exploited in the new approach described here to solve the problem of memory complexity.

Some modern FPGAs, like the Xilinx' Virtex family, use LUTs to implement logic functions [10]. Powerful synthesis tools like the Xilinx' Foundation® suite, aid the designers to efficiently map their applications onto the FPGAs. In order to use as few FPGA resources as possible, synthesis tools analyse designs and minimise the logic needed to implement them.

It is possible to simplify the memory on every MUXTREE cell by defining a LUT that represent the genome and then allowing the synthesiser optimise it. Since not all the configuration bits are used in all the cells, it is possible for the synthesiser to literally eliminate bits of memory. For example, cells that are not used must be programmed to a default value, resulting in many cells having the same genome. The synthesiser is able to detect configuration bits that will not change under any circumstance during the operation of the embryonic array. To save FPGA resources, these bits are tied to a logic value instead of generate them using logic. Such optimisation techniques generate architectures suitable for implementation in an FPGA. A simple example will now be discussed, followed by its implementation.

4   Example

To investigate the efficiency of the new approach for designing the MUXTREE's genome memory, the design of a programmable frequency divider is presented. Figure 2 shows the circuit's block diagram. This circuit is an improved version of the frequency divider presented in [5]. It consists of a 3-bit selector that latches either a constant **n** (the division factor), or the next state of a 3-bit down counter. The selector is controlled by the output of a zero detector. The circuit generates a 1-cycle low-pulse when the down counter reaches the 000 state. The output of the zero-detector will have a frequency lower than **F** by a factor proportional to **n**.

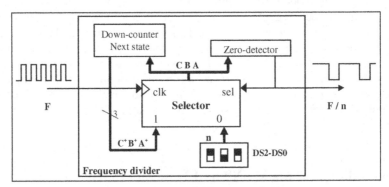

Fig. 2 Programmable frequency divider

Figure 3 shows the implementation of circuit in figure 2 using multiplexers.

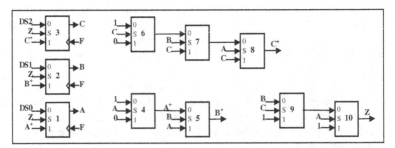

Fig. 3 Frequency divider implemented using multiplexers

In figure 3, every multiplexer corresponds to a node in the corresponding Ordered Binary Decision Diagram (OBDD) [3]. **A**, **B** and **C** are the outputs of the 3-bit down counter, **C** being the most significant bit. Multiplexers 1, 2 and 3 update their outputs on the rising edge of **F**; they implement the selector block. **DS2**, **DS1** and **DS0** are used to set the value of **n**.

Figure 4 shows the circuits in figure 3 mapped into an embryonic array. The numbers on each cell correspond with the numbers assigned to the multiplexers. Cells labelled **S** are spare cells. Cells labelled **R** are routing cells. Routing cells are needed to propagate information between non-neighbouring cells.

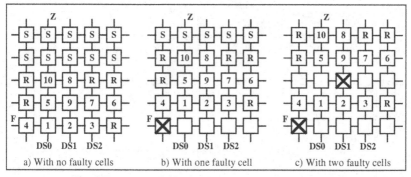

Fig. 4 Frequency divider implemented in embryonic array

At the present stage of the project, the mapping {logic equations}→{OBDDs}→ {Multiplexers}→{Embryonic array}, is done manually. An automated mechanism will be needed to implement applications that require a large number of multiplexers.

Figure 5 shows the content of the configuration register that configures each cell.

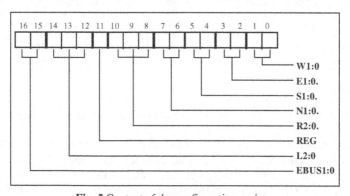

Fig. 5 Content of the configuration register

- **EBUS1:0-** Selects the selection input for the main multiplexer in every cell.
- **L2:0, R2:0-** Select the left (**L**) and right (**R**) inputs of the main multiplexer. For each input, one of eight possible signals is chosen.
- **REG-** If this bit is 1, the output of the cell becomes the registered output of the main multiplexer. If it is 0, the direct non-registered output is selected.
- **N1:0, E1:0, W1:0, S1:0-** These bit-pairs control the outputs of the I/O router.

Figure 6 shows the VHDL code of a look-up table with eight inputs (LUT-8) containing the genome that implements the frequency divider. Note that most of the 64 combinations presented in the inputs generate the code for the transparent configuration (0A200h). The synthesiser eliminates all the redundant bits, resulting in a very compact LUT that performs the function of the genome memory. Spare cells have the same configuration as transparent and routing cells; i.e. they just propagate signals across them.

```
library IEEE;
use IEEE.std_logic_1164.all;

entity Mem_freq_div is
  port (ok: in STD_LOGIC; -- OK status signal
        xy: in STD_LOGIC_VECTOR (5 downto 0); -- Co-ordinates
        conf: out STD_LOGIC_VECTOR (16 downto 0)); -- Genes
end Mem_freq_div;

architecture Mem_freq_div_arch of Mem_freq_div is
type REG is array (16 downto 0) of bit;
begin
  process(ok,xy)
    begin
      if (ok = '0') then
          conf <= "01010001000000000"; --0A200(Transparent)
      else
        case xy i
          when "000000" => conf <= "01000000100001100"; --0810C
          when "000001" => conf <= "00111111011010111"; --07ED7
          when "000010" => conf <= "00111111000000000"; --07E00
          when "000011" => conf <= "00111111011100000"; --07EE0
          when "000100" => conf <= "01010001000000001"; --0A201
          when "001001" => conf <= "10010010000001100"; --1240C
          when "001010" => conf <= "10001001001010100"; --11254
          when "001011" => conf <= "11010011100100010"; --1A722
          when "001100" => conf <= "11000000100010011"; --18113
          when "010001" => conf <= "00001001111111000"; --013F8
          when "010010" => conf <= "00110001100110000"; --06330
          when others    => conf <= "01010001000000000"; --0A200
        end case;
      end if;
    end process;
end Mem_freq_div_arch;
```

Fig. 6 VHDL code of genome to implement the frequency divider in figure 2

Schematic capture, VHDL synthesis and simulations were done using the Xilinx Foundation suite. Figure 7 shows the simulation results obtained for the frequency divider. Signals labelled **CONFAA** show the value of the configuration register in cell [0,0], signals **CONFAB** correspond to cell [0,1], and so on. Signal OKAA group together the lines that control the status of all the cells. A logic zero in one of this lines forces a fault in one of the cells. **CLKI** is the system clock and also the frequency being divided by the circuit in the example. Signals in bus **SIBUSAB** set the value of **n**, the division factor. **U17.NOBUS** is the output of the frequency divider.

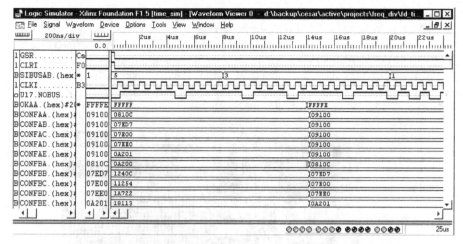

Fig. 7 Time simulation of frequency divider in figure 2

Simulation results in figure 7 show cell [0,0] failing at 14 μs. When the fault occurs, row **A** becomes transparent and row **B** takes its place, signalled by the apparent shift of configuration registers from one row to the other. After a reconfiguration period, the frequency divider keeps providing its function.

Figure 8 shows in detail the process of reconfiguration taking place.

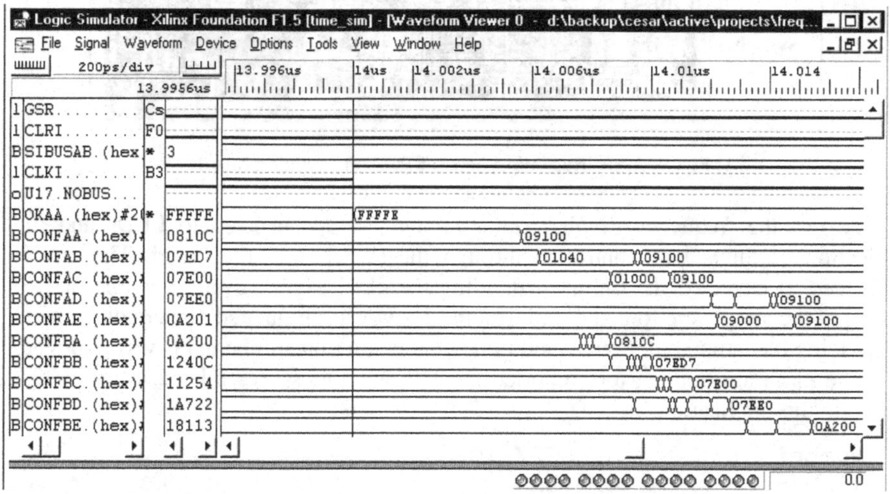

Fig. 8 Detail of reconfiguration process taking place after the injection of a fault

The process of reconfiguration comprises the following chain of events: {propagation of status signals on the row being eliminated}→{re-calculation of co-ordinates}→{decoding of co-ordinates to generate the new configuration word}→ {system stabilisation}. Figure 8 shows that, for this particular example, the process of reconfiguration takes 15.4 ns. Time to reconfiguration is size-of-array dependant.

Once validated by simulation, the circuit was downloaded to a prototyping board called Virtual Workbench®, developed by Virtual Computer Corporation™ [11]. A Virtual Workbench board supports various features of the Virtex FPGA, including a Flash, SRAM, SDRAM, temperature monitoring, communications and connectors for daughter cards. Also provides input and output devices like push-buttons, DIP-switches and LEDs, connected directly to the pins of a Virtex XCV300 device. Figure 9 shows a photograph of the Virtual Workbench board.

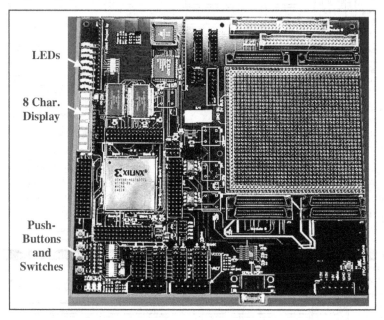

Fig. 9 The Virtual Workbench FPGA prototyping board

To test the frequency divider, status signals were connected to LEDs and the injection of faults was controlled through the switches in the board. The circuit behaved as expected. It continued dividing the clock signal, even after one or two rows of the array were forced to fail. If a third row was eliminated, the circuit stopped providing its output, but it recovered immediately after one of the faults was removed.

The place and routing tool reported the following messages related to the FPGA used in this application:

```
Number of external GCLK IOBs        1 out of 4        25%
Number of external IOBs             107 out of 260    41%
Number of SLICES                    526 out of 3072   17%
```

From these messages, it may be concluded that FPGA resources are still available to implement either more cells or additional logic to support the function of the array.

5   Conclusions and future work

Embryonics proposes a new family of FPGA devices inspired by the ontogenetic process of multicellular organisms. As such, the ultimate goal of the embryonics project is to provide a new family of integrated circuits. However, the high costs associated with circuit fabrication have prevented the widespread use of embryonics ideas.

This paper presented a new proposal for the design of the MUXTREE embryonic cell. It is shown that implementing the memory block as a look-up table and allowing the synthesis tools to optimise it, results in an area-efficient circuit, suitable for implementation in commercial LUT-based FPGAs. By implementing the genome as a LUT, it is possible to create hybrid systems where only part of the physical FPGA is used to implement the embryonic array. The remaining resources can be used to implement other conventional, non-embryonic functions.

The physical implementation of a frequency divider has confirmed that it is possible to integrate a complete 5×5 embryonic array in a commercial XCV300 Virtex FPGA.

For the group working in embryonics at the University of York, this has been the first physical implementation of embryonic arrays on a commercial FPGA. Future work will explore the capabilities of the synthesis tools in order to determine the maximum number of cells that can be fitted in one XCV300 device.

The results obtained so far demonstrate that the embryonics concept can be used today, with no need to wait for the specifically designed "embryonic FPGA". However, it is expected that pure embryonic circuits will play an important role in the future of fault-tolerant design.

Acknowledgements

This work has been partially supported by Mexico's Government under grants CONACYT-111183 and IIE-9611310226.

References

1. Mange D. and Tomassini M. (Eds.), Bio-Inspired Computing Machines, Presses Polytechniques et Universitaires Romandes, Switzerland, 1998
2. Marchal P., Invited presentation at the 1st NASA/DoD Workshop on Evolvable Hardware, Pasadena, USA, July 1999, http://cism.jpl.nasa.gov/events/nasa_eh/papers/Presentation2.ppt
3. Ortega C. and Tyrrell A., "Biologically Inspired Fault-Tolerant Architectures for Real-Time Control Applications", Control Engineering Practice, July 1999, pp. 673-678
4. Ortega C. and Tyrrell A., "Fault-tolerant Systems: The way Biology does it!", Proceedings Euromicro 97, Budapest, IEEE CS Press, September 1997, pp.146-151
5. Ortega C. and Tyrrell A., "MUXTREE revisited: Embryonics as a Reconfiguration Strategy in Fault-Tolerant Processor Arrays", Lecture Notes in Computer Science 1478, Springer-Verlag, 1998, pp.206-217

6. Ortega C. and Tyrrell A., "Reliability Analysis in Self-Repairing Embryonic Systems", in Stoica A. et al. (Eds.), Procs. of 1st NASA/DoD Workshop on Evolvable Hardware, Pasadena, CA, USA, IEEE Computer Society, July 1999, pp.120-128
7. Tempesti G., "A Robust Multiplexer-based FPGA Inspired by Biological Systems", Special Issue of JSA on Dependable Parallel Computer Systems, Feb. 1997, pp.719-733
8. Wolpert L., The Triumph of the Embryo, Oxford University Press, 1991
9. Ortega C. and Tyrrell A., "Design of a Basic Cell to Construct Embryonic Arrays", IEE Procs. on Computers and Digital Techniques, Vol.145-3, May, 1998, pp.242-248
10. Xilinx, Virtex™ 2.5V FPGA Advance Product Specification, version 1.6, July 1999
11. Virtual Computer Corporation, The Virtual Workbench Guide, version 1.01, 1999

Everything on the Chip: A Hardware-Based Self-Contained Spatially-Structured Genetic Algorithm for Signal Processing

Simon Perkins, Reid Porter, and Neal Harvey

Los Alamos National Laboratories, Los Alamos, NM 87545, USA,
{s.perkins,rporter,harve}@lanl.gov,
WWW home pages: http://nis-www.lanl.gov/{s̃imes,r̃porter,h̃arve}

Abstract. Evolutionary algorithms are useful optimization tools but are very time consuming to run. We present a self-contained FPGA-based implementation of a spatially-structured evolutionary algorithm that provides significant speedup over conventional serial processing in three ways: (a) efficient hardware-pipelined fitness evaluation of individuals, (b) evaluation of an entire population of individuals in parallel, and (c) elimination of slow off-chip communication. We demonstrate using the system to solve a non-trivial signal reconstruction problem using a non-linear digital filter on a Xilinx Virtex FPGA, and find a speedup factor of over 1000 compared to a C implementation of the same system. The general principles behind the system are very scalable, and as FPGAs become even larger in the future, similar systems will provide extremely large speedups over serial processing.

1   Introduction

Evolutionary algorithms (EAs) are perhaps the most general purpose practical optimization technique in use today. The same basic processes of evaluation, selection and recombination can be applied to any problem for which a fitness function and representation can be defined. The price we pay for this generality is that EAs typically take a very long time to find a solution for hard problems, as compared to more problem-specific techniques that make use of information about the nature of the problem being tackled.

The core process of almost all EAs involves performing a very large number of fitness evaluations. As a result, any way of speeding up fitness evaluations has significant consequences for how long the EA will take to solve a problem. Many researchers have looked at speeding up evaluation using reconfigurable hardware such as FPGAs (see [Higuchi et al., 1996] and [Sipper et al., 1998] for examples). Hardware fitness evaluation can provide a large speedup over software fitness evaluation in those cases where the fitness evaluation can be decomposed into many simple steps that can be carried out in parallel on the chip. For example, image processing algorithms are very time consuming to evaluate on a serial processor, but can be much more efficiently executed using a hardware pipeline.

J. Miller et al. (Eds.): ICES 2000, LNCS 1801, pp. 165–174, 2000.

A complementary approach to speeding up fitness evaluation is to use the hardware to carry out many fitness evaluations in parallel. In the limit, this would involve putting an entire EA population onto an FPGA. However, putting many evaluation units on a chip pushes very hard against the space limits of current FPGAs and so there are very few examples of this technique in the literature. One active area where we do see whole populations on a chip is the 'cellular programming' paradigm [Sipper, 1997]. CP solves the problem of fitting many units on a chip by making those units very simple — the basic individual in a CP system is a single cell of a cellular automaton, which implements only a very simple combinational logic function of its neighbors' states. The idea is that, although each component is very simple, complex large-scale behaviour can be achieved through local interactions between individuals in the evolving population. While this approach is very promising, it seems clear that it will not work for all problems. For more general EAs, we would like each individual to be capable of an arbitrary amount of computation by itself. Unfortunately, more complex individuals require more space per individual and space has always been at a premium on FPGAs.

However, the state of the art has recently taken a major leap forward with the arrival of the Xilinx Virtex FPGA [Xilinx, 1999]: a 'million-gate equivalent' part with a vast array of exciting hardware features that may well make it ideally suited to EA implementations. With the arrival of these chips it has now become feasible to place a reasonably sized population of reasonably complex individuals on a single FPGA part, with the following important consequence: if we choose our problem and our individuals appropriately we can then use the hardware to get a speedup in three significant ways:

- We get speedup by evaluating many individuals in parallel.
- We get speedup because with the extra space we can perform each of those evaluations in efficient hardware-parallel fashion.
- By putting the entire EA onto the chip, we get speedup by eliminating relatively slow chip-to-host communication and/or FPGA reconfiguration.

In this paper, we present preliminary results evolving a population of non-linear digital filters on a single Virtex chip to solve a non-trivial 1-D signal reconstruction problem. More important than the particular application though are the general architectural principles which can be expanded as FPGAs grow larger to provide greater and greater speedups compared to conventional processing.

2   Problem Definition and General Approach

2.1   1-D Signal Reconstruction

In the experiments reported in this paper, we look at the problem of reconstructing a pure signal from a digitized signal that has been corrupted with 'shot' noise: upsets that occur randomly at a constant expected rate and that set the value of a sample to a totally random level. Such reconstruction problems occur in many practical situations, such as cleaning up signals sent over noisy channels in telecommunications systems. We would like our EA to design a filter that transforms the corrupted signal into a close approximation of the original one.

Stack Filters One important class of techniques for performing reconstruction makes use of 'stack filters'. A stack filter (SF) is a sliding window non-linear filter whose output at each window position is determined by applying a particular *positive* boolean function[1] (PBF) to a 'threshold decomposed' representation of the window.

The threshold decomposition process and subsequent filter operation is easily visualized as follows: take the 1-D string of values in the window to which the SF is being applied, and imagine them forming a 2-D 'wall' where the height of the wall at each window location corresponds to the value in that location. Then imagine taking 1-D horizontal slices of the wall, at every possible level (since we assume the signal values are discrete, there are only a finite number of such slices). Each slice gives us a 1-D string of boolean values where 'true' corresponds to 'inside the wall' and 'false' corresponds to 'outside the wall'. We apply the filter's PBF to each of these slices and find the total number of 'true' results we got from all the slices. This number is then the output of the filter at that window location. The fact that we use a positive boolean function, ensures that the output value is one of the input values.

Stack filters include as a subset such commonly used non-linear filters as the median filter, weighted median filter and other rank-order filters.

A number of researchers have used genetic algorithms to optimize stack filters, notably Chu in [Chu, 1990]. At least one researcher has also looked at evolving stack filters on an FPGA [Woolfries et al., 1998], but using the more 'conventional' technique of downloading and evaluating them one at a time on the chip.

Implementing Stack Filters in Hardware The threshold decomposition phase of stack filters can be time-consuming. For an 8-bit signal, there are 256 different levels at which to threshold, and hence each filter operation requires 256 evaluations of the PBF. Fortunately, there are more efficient ways of computing the same result. The technique we use was proposed by Chen [Chen, 1989] specifically for implementation in hardware. It relies on binary search and only requires a number of PBF evaluations equal to the number of bits used to represent the signal. The method assumes that the input values to the filter arrive as parallel bit-streams, MSB first. The PBF is first applied to the most significant bits of the input values, giving the MSB of the output value. If any of the input bits have a different value to the output bit, then that input stream is 'latched'

---

[1] A positive boolean function is a boolean function that can be written using just AND and OR and without negating any of the inputs.

at its current value. The PBF is then applied to the next bits from the streams, giving the next output bit, and the process repeats.

Stick Filters... Chen's method produces results equivalent to the proper stack filter algorithm, if the boolean function used is a positive boolean function. However, in the context of a genetic algorithm, it turns out to be fairly inconvenient to generate PBFs. The reason is that only a small fraction of all possible boolean functions for a given number of inputs are actually positive boolean functions, and it is time-consuming to check whether any particular boolean function is positive or not. In our experiments, we use a 5-element window, and so the boolean function can be defined by a truth table containing 32 1-bit values. There are $2^{32} \approx 4.3 \times 10^9$ possible such truth tables but in fact only 7581 of these represent positive boolean functions.

Our solution is simply to ignore the problem. We apply Chen's method using *arbitrary* boolean functions. The result is a stack filter if the boolean function happens to be one of the 7581 5-input PBFs, and 'something else' if it isn't. Exactly 'what' is difficult to describe concisely. The resulting filters are a rather strange class of non-linear digital filters that are a superset of stack filters. We call them 'stick filters', in reference to the way in which input bit stream values get 'stuck' if they disagree with the output value. The weird and wonderful space of stick filters is ideally suited to exploration by a genetic algorithm.

2.2   Genetic Algorithm Details

Representation Once the window size has been fixed, the operation of a stick filter is defined by its boolean function. For the 5-element windows we use in our experiments, an arbitrary boolean function can be represented by a 32 element truth table. It seems sensible to use a direct genomic representation here, and so the genome for an individual in our GA is simply a binary string giving the truth table for its boolean function, with the output value for an input of '00000' at the left end. Genetic operations are carried out directly on this representation.

Evolutionary Algorithm We use a fine-grained spatially-structured genetic algorithm that is similar to that used by Moshe Sipper in his cellular programming work [Sipper, 1997] and to other more conventional GA practitioners such as [Manderick and Spiessens, 1989]. A population of 48 evolutionary cells is distributed over an 6 × 8 grid on the FPGA. The cells are initialized with random truth tables. Each cell also maintains an error counter E, which is initialized to zero.

Once initialization is complete, evaluation begins. Each cell in the grid receives the same training data, consisting of a corrupted signal S^c and an uncorrupted version of the signal S^u, with the latter delayed by 2 time steps with respect to the former. At each time step (after the first 4), it applies the stick filter to a window consisting of the last 5 samples of S^c to generate a 'reconstructed' signal S^r. The absolute difference d between this output and the uncorrupted signal's value at that time is computed and this value is added to E. After a fixed number of times steps τ, fitness evaluation is complete. This process is illustrated in Figure 1.

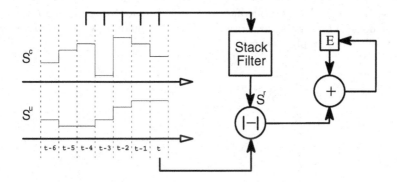

Fig. 1. Basic step of stack filter evaluation phase.

Once each cell's total error E has been determined, the GA goes into a breeding mode. Each cell compares its own error with that of each of its four neighbors to the north, east, south and west on the grid (toroidal wraparound is used at the edges). Once the fittest cell in its neighborhood has been found (this can be itself), uniform crossover is performed with the cell and its fittest neighbor. Note that if a cell is the fittest in its neighborhood then this uniform crossover leaves the cell unchanged. Point mutation is then applied to every element in the truth table of each cell with probability p_m. The effect of mutation is to flip that element's value.

Once breeding is over, the error counter is reset to zero, and the cycle repeats.

3   Simulation Experiments

3.1   Experimental Setup

A software simulator was written in Java, to test out the ideas behind the project. This functioned in an identical manner to that described above (except for running on a serial computer rather than an FPGA!). A signal consisting of a sine-wave with period equal to 20 time steps, quantized into 256 levels and corrupted by adding random shot noise was used as training input. Noise was added at an average rate (Poisson distributed) of 0.2 per time step. The effect of the noise was to drive the signal to a random value between 0 and 255, for that time step only.

The training period T was set to 2048 time steps, and the probability of a bit mutation p_m was set to 0.01 per bit.

3.2   Results

Figure 2 shows the way in which the error of the best individual in the population
fell with generation number. The y-axis indicates the average error per sample
of the best individual. Also shown is the error obtained on the same signal data,
using a conventional 5-input median filter — a commonly used filter for removing
impulse noise. The curves were obtained by averaging over five separate runs with
different random number seeds.

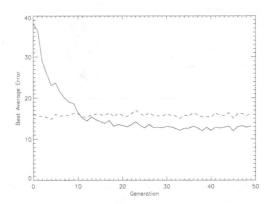

Fig. 2. Best average error vs. time graphs. The solid line shows the results from the
best evolved filter, while the dashed line shows the results for a conventional 5-input
median filter.

The graphs show clearly that the best of the initial random population fails to
do as well the median filter, but after about 10 generations the best of the evolved
filters begin to do consistently better than the median filter. There is some
variation in the error scores from one generation to the next due to variations
in the randomly generated input signals, but even so, the evolved filter always
does better than the median.

A possible objection to this conclusion is that the 'best' evolved filter is
chosen from 48 different candidates every generation, which could lead to an
unfair bias in favor of the evolved filters. To counter this objection we took the
most commonly evolved final genome (see below) and compared it head-to-head
against the median filter on a fresh corrupted signal for an evaluation period of
30000 cycles. We also tested the filters on signals with different periods and different noise rates. Two-tailed paired sample t-tests were used to investigate (and
confidently reject) the null hypothesis that the filters were performing identically
on average. Table 1 summarizes these results.

An interesting fact is that in all the five runs performed for Figure 2, the same
individual appeared most frequently in the final generation. This individual had
the genome: 000000110001111110000011100111111.

Signal Period	Noise Rate	Median Filter	Evolved Filter	p
20	0.2	15.9	12.9	$\ll 0.001$
20	0.4	32.5	29.7	$\ll 0.001$
10	0.2	36.1	19.8	$\ll 0.001$
10	0.4	50.6	38.4	$\ll 0.001$

Table 1. Comparison of the best evolved filter with the median filter on various test problems. The third and fourth columns give the mean error per sample achieved during the tests. The final column gives the p-value for the null hypothesis that the mean performance of the filters is the same.

On closer inspection this turns out to be a positive boolean function (despite the fact that there was no explicit attempt to produce one) and seems to perform a 'weighted-center' median filter.

4   FPGA Implementation

Our design is targeted at the Annapolis Microsystems WildCard. This PCMCIA card contains a Xilinx Virtex 300 part and two independent banks of 256kB SRAM. The top level architecture for a cell is illustrated in Figure 3.

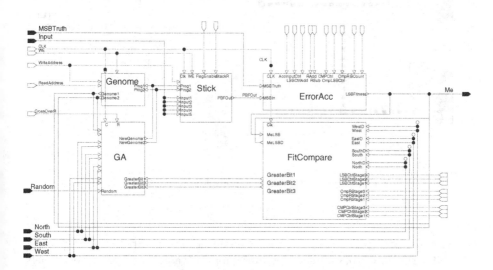

Fig. 3. Top level FPGA design.

At any one time a cell is in one of three modes. The mode is determined by control lines shown in Figure reffig:TopLevel with unshaded terminal nodes. Since all cells on the array are operating synchronously these control lines are driven by a central controller common to all cells. A cell receives both genome and fitness information from its four neighbors illustrated by the shaded terminals 'North', 'South', 'East' and 'West'. The cell's own genome and fitness information are communicated to its four neighbors via the black terminal 'Me'.

In the first mode of operation, the stick filter's fitness is evaluated over 2048 input samples within the 'ErrorAcc' function block.

In the second mode of operation, this accumulated error is compared to error values of the cell's neighbors in the 'FitCompare' function block. The fittest neighbor is determined, and communicated to the 'GA' function block through the 'GreaterBit' control lines.

In the third mode of operation, the 'GA' function block makes use of a pre-loaded random bit stream entering through the 'Random' terminal to implement uniform crossover and mutation as described before, and to reprogram the stick filter.

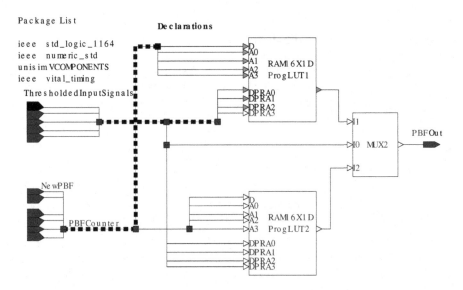

Fig. 4. Details of PBF implementation.

The core of a cell's signal processing power lies in the 'Stick' function block. The bit-serial nature of Chen's stack filter implementation results in an extremely compact architecture that is well suited to the Virtex FPGA. The complete filter can be implemented in just 3 Virtex CLBs (Configurable Logic Blocks). 1.5 of these CLBs are used to implement the cell's boolean function in the manner shown in Figure 4. The two programmable LUTs (configured as dual port RAMS) receive 4 of the 5 input samples through the 'ReadAddress' terminal and output to a multiplexer controlled by the 5th input sample. This configuration can implement any function of five variables. A very similar configuration is found in the 'Genome' function block which holds a duplicate copy of the stick filter's 5-input boolean logic function.

Training data is pre-loaded into onboard memory in bit serial order, most significant bit first. On reset the cell array remains in an 'idle' configuration. At this time each cell's genome is configurable by the host processor. The array is activated when the host processor writes the number of desired generations to an on-chip register. The array then iterates for this number of generations, and, when complete, initiates an interrupt and returns to the idle configuration. At this stage the host program may once again read and write cell genomes.

A VHDL model of the GA has been both simulated and synthesized success-fully for the Virtex 300 FPGA used on the WildCard. Table 2 analyzes the area requirements for the various components. Each CLB slice contains two 4-input look-up tables and two 1 bit registers. The controller moves each cell through its three modes of operation and also provides synchronization mechanisms for communicating with the host. The WildCard interfaces enable communication between the FPGA and onboard memory as well as to the host through the 32 bit CardBus.

Component	Number of CLB Slices	Percentage of Total
48 Cell Array	1904	62
Controller	43	1.5
WildCard Interfaces	549	18
Total	2496	81.5

Table 2. Space utilization on the Virtex 300.

Timing Experiments

For 8 bits/sample data, cells require 16 clock cycles to process one sample. Due to high fan out in synchronizing the cell array, clock rates were restricted to 20MHz, however higher rates are believed to be obtainable with more detailed design work. At this clock speed, the FPGA can run through one of the above experiments, using 48 individuals, run for 50 generations, and evaluated for 2048 time steps each generation, in an impressive 1.65 ms! In contrast, the Java simulator takes well over 5 minutes to achieve the same result.

Of course, the Java simulator was optimized for visualization rather than speed, so the core of the GA was re-implemented using C. The optimized C code took 1.8 s to perform the above experiment — still over 1000 times slower than the FPGA.

5   Further Work and Conclusions

At the time of writing the synthesized design has not actually been placed onto a real FPGA due to problems with our hardware, but we hope to remedy this in the near future.

It should be pointed out that the problem tackled here, while non-trivial, is not all that hard either — after all, a software version of the GA was able to find good solutions in just a few seconds! In order to demonstrate a real advantage we would like to extend the work here to look at optimizing more complicated types of signal and image processing functions that are intractably difficult to optimize in software. As an example, merely increasing the window neighborhood of our stick filter from 5 to 7 elements, increases the number of potentially useful positive boolean functions from 7851 to over 3×10^{10}! We can also imagine using FPGAs in situations where relatively simple problems such as this one must be solved extremely quickly. For instance, suppose we want to install an adaptive signal reconstructor on a communications channel that must re-evolve to match changing noise conditions (using a known signal sent over the channel as truth data), every few minutes.

As FPGAs continue to grow in size it will become easier and easier to fit larger and larger populations of more complex individuals on single chips and so we confidently expect this area of research to be a fruitful one.

References

[Chen, 1989] Chen, K. (1989). Bit-serial realizations of a class of nonlinear filters based on positive boolean functions. *IEEE Transactions on Circuits and Systems*, 36(6).

[Chu, 1990] Chu, C. (1990). The application of an adaptive plan to the configuration of nonlinear image processing algorithms. In *SPIE Proceedings - Nonlinear Image Processing*, volume 1247, pages 248–257.

[Higuchi et al., 1996] Higuchi, T., Iwata, M., and Liu, W., editors (1996). *Evolvable Systems: From Biology to Hardware: Proc. ICES '96*, volume 1259 of *Lecture Notes in Computer Science*. Springer.

[Manderick and Spiessens, 1989] Manderick, B. and Spiessens, P. (1989). Fine-grained parallel genetic algorithms. In Schaffer, J., editor, *Proc. 3rd Int. Conf on Genetic Algorithms*. Morgan Kaufmann.

[Sipper, 1997] Sipper, M. (1997). *Evolution of Parallel Cellular Machines*, volume 1194 of *Lecture Notes in Computer Science*. Springer-Verlag.

[Sipper et al., 1998] Sipper, M., Mange, D., and Pérez-Uribe, A., editors (1998). *Evolvable Systems: From Biology to Hardware: Proc. ICES '98*, volume 1478 of *Lecture Notes in Computer Science*. Springer.

[Woolfries et al., 1998] Woolfries, N., Lysaght, P., Marshall, P., McGregor, S., and Robinson, G. (1998). Fast adaptive image processing in fpgas using stack filters. In Hartenstein, R. W. and Keevallik, A., editors, *Field Programmable Logic and Applications: From FPGAs to Computing Paradigm: 8th Int. Workshop*.

[Xilinx, 1999] Xilinx, I. (1999). Virtex 2.5v field programmable gate arrays. Advance Product Specification. Version 1.3.

Evolutionary Techniques in Physical Robotics

Jordan B. Pollack, Hod Lipson, Sevan Ficici, Pablo Funes, Greg Hornby, and
Richard A. Watson

Computer Science Dept., Brandeis University, Waltham, MA 02454, USA,
pollack@cs.brandeis.edu,
http://www.demo.cs.brandeis.edu

Abstract. Evolutionary and coevolutionary techniques have become a
popular area of research for those interested in automated design. One of
the cutting edge issues in this field is the ability to apply these techniques
to real physical systems with all the complexities and affordances that
such systems present. Here we present a selection of our work each of
which advances the richness of the evolutionary substrate in one or more
dimensions. We overview research in four areas: a) High part-count static
structures that are buildable, b) The use of commercial CAD/CAM systems as a simulated substrate, c) Dynamic electromechanical systems
with complex morphology that can be built automatically, and d) Evolutionary techniques distributed in a physical population of robots.

1 Introduction

The field of Robotics today faces a practical problem: most problems in the
physical world are too difficult for the current state of the art. The difficulties
associated with designing, building and controlling robots have led to a stasis [1] and robots in industry are only applied to simple and highly repetitive
manufacturing tasks.

Even though sophisticated teleoperated machines with sensors and actuators
have found important applications (exploration of inaccessible environments for
example), they leave very little decision, if at all, to the on-board software [2].
Hopes for autonomous robotics were raised high several times in the past — the
early days of AI brought us Shakey the robot [3], and the eighties the "subsump-
tion architecture" [4]. Yet today, fully autonomous robots with minds of their
own are not a reality.

The central issue we begin to address is how to get a higher level of complex
physicality under control with less human design cost. We seek more controlled
and moving mechanical parts, more sensors, more nonlinear interacting degrees
of freedom — without entailing both the huge costs of human design, programming, manufacture and operation. We suggest that this can be achieved only
when robot design and construction are fully automatic and the constructions
inexpensive enough to be disposable and/or recyclable.

The focus of our research is thus how to automate the integrated design of
bodies and brains using a coevolutionary learning approach. Brain/body coevo-

lution is a popular idea, but evolution of robot bodies is usually restricted to adjusting a few morphological parameters in an otherwise fixed, human-engineered automaton [5–7]. We propose that the key is to evolve both the brain and the body, simultaneously and continuously, from a simple controllable mechanism to one of sufficient complexity for a task. We then require a replication process, that brings an exact copy of the evolved machine into reality. Finally, the transfer between a simulated environment and reality — the "reality gap" — needs further embodied adaptation.

We see three technologies that are maturing past threshold to make this possible. One is the increasing fidelity of advanced mechanical design simulation, stimulated by profits from successful software competition [8]. The second is rapid, one-off prototyping and manufacture, which is proceeding from 3D plastic layering to stronger composite and metal (sintering) technology [9]. The third is our understanding of coevolutionary machine learning in design and intelligent control of complex systems [10–12].

2   Coevolution

Coevolution, when successful, dynamically creates a series of learning environments each slightly more complex than the last, and a series of learners which are tuned to adapt in those environments. Sims' work [13] on body-brain coevolution and the more recent Framsticks simulator [14] demonstrated that the neural controllers and simulated bodies could be coevolved. The goal of our research in coevolutionary robotics is to replicate and extend results from virtual simulations like these to the reality of computer designed and constructed special-purpose machines that can adapt to real environments.

We are working on coevolutionary algorithms to develop control programs operating realistic physical device simulators, both commercial-off-the-shelf and our own custom simulators, where we finish the evolution inside real embodied robots. We are ultimately interested in mechanical structures that have complex physicality of more degrees of freedom than anything that has ever been controlled by human designed algorithms, with lower engineering costs than currently possible because of minimal human design involvement in the product.

It is not feasible that controllers for complete structures could be evolved (in simulation or otherwise) without first evolving controllers for simpler constructions. Compared to the traditional form of evolutionary robotics [15–19] which serially downloads controllers into a piece of hardware, it is relatively easy to explore the space of body constructions in simulation. Realistic simulation is also crucial for providing a rich and nonlinear universe. However, while simulation creates the ability to explore the space of constructions far faster than real-world building and evaluation could, transfer to real constructions is problematic. Because of the complex emergent interactions between a machine and its environment, learning and readaptation must occur in "embodied" form as well [20–22].

3 Research Thrusts

We have four major thrusts in achieving Fully Automated Design (FAD) and manufacture of high-parts-count autonomous robots. The first is evolution inside simulation, but in simulations more and more realistic so the results are not simply visually believable, as in Sims' work, but also tie into manufacturing processes. Indeed, interfacing evolutionary computation systems to commercial off-the-shelf CAD/CAM systems through developer interfaces to mechanical simulation programs seems as restrictive as developing programming languages for 8K memory microcomputers in the middle 1970's. However, even though the current mechanical simulation packages are "advisory" rather than blue-print generating, and are less efficient than research code, as computer power grows and computer-integrated manufacturing expands, these highly capitalized software products will absorb and surpass research code, and moreover will stay current with the emerging interfaces to future digital factories. The second thrust is to evolve automatically-buildable machines, using custom simulation programs. Here, we are willing to reduce the universe of mechanisms we are working with in order to increase the fidelity and efficiency of the simulators and reduce the cost of building resulting machines. The third is to perform evolution directly inside real hardware, which escapes the known limitations of simulation and defines a technology supporting the final learning in embodied form. This is perhaps the hardest task because of the power, communication and other reality constraints. The fourth thrust addresses handling high part-count structures with realistically complexity. We have preliminary and promising results in each of these four areas, which we outline below.

3.1 Evolution in Simulation

We have been doing evolution of neural-network controllers inside realistic CAD simulations as a prelude to doing body reconfiguration and coevolution. Our lab is using CAD/CAM software package that comprises a feature based solid-modeling system [1]. Widely used in industry, it includes a mechanical simulation component that can simulate the function of real-world mechanisms, including gears, latches, cams and stops. This program has a fully articulated development interface to the C programming language, which we have used in order to interface its models to our evolutionary recurrent neural network software.

To date, we have used this system with evolved recurrent neural controllers for one and two segment inverted pendulums and for Luxo (an animated lamp creature, figure 1). Many researchers have evolved such controllers in simulation, but no one has continuously deformed the simulation and brought the evolved controllers along, and no one else has achieved neural control inside commercial simulations. We believe this should lead to easy replication, extension, and transfer of our work.

---

[1] Parametric Technology Corporation's *Pro/Engineer*.

We have successful initial experiments consisting of evolving recurrent neural network controllers for the double-pole balancing problem, where we slowly "morphed" the body simulator by simulating a stiff spring at the joint connecting the two poles and relaxing its stiffness.

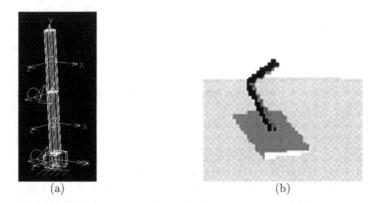

(a)   (b)

Fig. 1. Commercial CAD models for which we evolved recurrent neural net controllers: (a) a two-segment inverted pendulum (b) A Luxo lamp.

Some of the ways to achieve continuous body deformation are:

- New links can be introduced with "no-op" control elements.
- The mass of new links can initially be very small and then incremented.
- The range of a joint can be small and then given greater freedom.
- A spring can be simulated at a joint and the spring constant relaxed.
- Gravity and other external load forces can be simulated lightly and then increased.

3.2   Buildable Simulation

Commercial CAD models are in fact not constrained enough to be buildable, because they assume a human provides numerous constraints to describe reality. In order to evolve both the morphology and behavior of autonomous mechanical devices that can be built, one must have a simulator that operates under many constraints, and a resultant controller that is adaptive enough to cover the gap between the simulated and real world. Features of a simulator for evolving morphology are:

- Representation —- should cover a universal space of mechanisms.
- Conservative — because simulation is never perfect, it should preserve a margin of safety.
- Efficient — it should be quicker to test in simulation than through physical production and test.

- Buildable — results should be convertible from a simulation to a real object.

One approach is to custom-build a simulator for modular robotic components, and then evolve either centralized or distributed controllers for them. In advance of a modular simulator with dynamics, we recently built a simulator for (static) Lego bricks, and used very simple evolutionary algorithms to create complex Lego structures, which were then manually constructed [23–25].

Our model considers the union between two bricks as a rigid joint between the centers of mass of each one, located at the center of the actual area of contact between them. This joint has a measurable torque capacity. That is, more than a certain amount of force applied at a certain distance from the joint will break the two bricks apart. The fundamental assumption of our model is this idealization of the union of two Lego bricks.

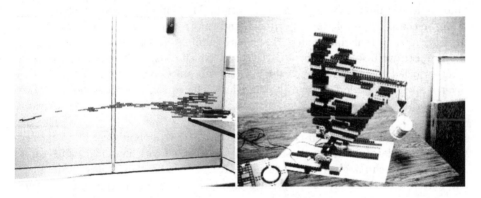

Fig. 2. Photographs of the FAD Lego Bridge (Cantilever) and Crane (Triangle). Photographs copyright Pablo Funes & Jordan Pollack, used by permission.

The genetic algorithm reliably builds structures that meet simple fitness goals, exploiting physical properties implicit in the simulation. Building the results of the evolutionary simulation (by hand) demonstrated the power and possibility of fully automated design. The long bridge of figure 2 shows that our simple system discovered the cantilever, while the weight-carrying crane shows it discovered the basic triangular support.

3.3   Evolution and construction of electromechanical systems

The next step is to add dynamics to modular buildable physical components. We are experimenting with a new process in which both robot morphology and control evolve in simulation and then replicate automatically into reality. The robots are comprised of only linear actuators and sigmoidal control neurons embodied in an arbitrary thermoplastic body. The entire configuration is evolved for a particular task and selected individuals are printed pre-assembled (except motors) using 3D solid printing (rapid prototyping) technology, later to be recycled into different forms. In doing so, we establish for the first time a complete physical evolution cycle. In this project, the evolutionary design approach assumes two main principles: (a) to minimize inductive bias, we must strive to use the lowest level building blocks possible, and (b) we coevolve the body and the control, so that that they stimulate and constrain each other.

We use arbitrary networks of linear actuators and bars for the morphology, and arbitrary networks of sigmoidal neurons for the control. Evolution is simulated starting with a soup of disconnected elements and continues over hundreds of generations of hundreds of machines, until creatures that are sufficiently proficient at the given task emerge. The simulator used in this research is based on quasi-static motion. The basic principle is that motion is broken down into a series of statically-stable frames solved independently. While quasi-static motion cannot describe high-momentum behavior such as jumping, it can accurately and rapidly simulate low-momentum motion. This kind of motion is sufficiently rich for the purpose of the experiment and, moreover, it is simple to induce in reality since all real-time control issues are eliminated [26, 27].

Several evolution runs were carried out for the task of locomotion. Fitness was awarded to machines according to the absolute average distance traveled over a specified period of neural activation. The evolved robots exhibited various methods of locomotion, including crawling, ratcheting and some forms of pedalism.

Selected robots are then replicated into reality: their bodies are first fleshed to accommodate motors and joints, and then copied into material using rapid prototyping technology. Temperature-controlled print head extrudes thermoplastic material layer by layer, so that the arbitrarily evolved morphology emerges pre-assembled as a solid three- dimensional structure without tooling or human intervention. Motors are then snapped in, and the evolved neural network is activated (figure 4). The robots then perform in reality as they did in simulation.

3.4   Embodied Evolution

Once a robot is built, it may well be necessary to "fine-tune" adaptation in the real world. Our approach is to perform adaptation via a decentralized evolutionary algorithm that is distributed and embodied within a population of robots. The distributed and asynchronous operation of this evolutionary method allows the potential for being scaled to very large populations of robots, on the order of hundreds or thousands, thus enabling speedup that is critical when using evolution in real robots.

Technologically, this introduces two main problems: long-term power and reprogramming [28,29]. Many robots' batteries last only for a few hours, and robots typically have to be attached to a PC for new programs to be uploaded. In order to do large group robot learning experiments, we have designed a continuous-power floor system, and utilized infra-red (IR) communications to transfer programs between robots. We are thus able to run a population of learning robots battery-free and wire-free for days at a time (figure 5). Evolution is not directed

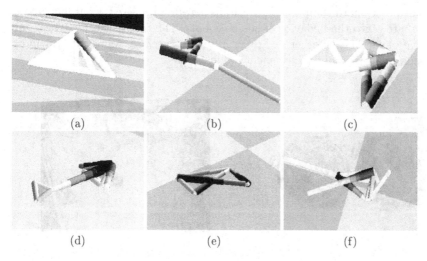

Fig. 3. (a) A tetrahedral mechanism that produces hinge-like motion and advances by pushing the central bar against the floor. (b) Bi-pedalism: the left and right limbs are advanced in alternating thrusts. (c) Moves its two articulated components to produce crab-like sideways motion. (d) While the upper two limbs push, the central body is retracted, and vice versa. (e) This simple mechanism uses the top bar to delicately shift balance from side to side, shifting the friction point to either side as it creates oscillatory motion and advances. (f) This mechanism has an elevated body, from which it pushes an actuator down directly onto the floor to create ratcheting motion. It has a few redundant bars dragged on the floor.

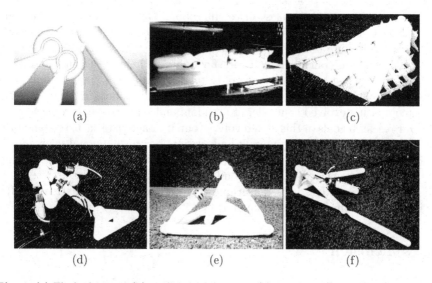

Fig. 4. (a) Fleshed joints, (b) replication progress, (c) pre-assembled robot (figure 3f), (d,e,f) final robots with assembled motors

by a central off-board computer that installs new programs to try out, but rather is distributed into the behavior of all the robots. The robots exchange program specifications with each other and this "culture" is used to 'reproduce' the more successful behaviors and achievement of local goals.

(a)   (b)

Fig. 5. 4" diameter robot picks up power from its environment and learns while on-line.

The control architecture is a simple neural network and the specifications for it are evolved online. Each robot tries parameters for the network and evaluates its own success. The more successful a robot is at the task, the more frequently it will broadcast its network specifications via its local-range IR communications channel. If another robot happens to be in range of the broadcast, it will adopt the broadcast value with a probability inversely related to its own success rate. Thus, successful robots attempt to influence others, and resist the influence of others, more frequently than less successful robots. We have shown this paradigm to be robust both in simulation and in real robots, allowing for the parallel, asynchronous evolution of large populations of robots with automatically developed controllers. These controllers compare favorably to human designs, and often surpass them when human designs fail to take all important environmental factors into account. Figure 6 shows averaged runs of the robots in a light seeking task, comparing evolved controllers to random and human-designed ones.

Our research goals in this area involve group interactive tasks based on multi-agent systems, such as group pursuit and evasion, box pushing, and team games. These domains have been out of reach of traditional evolutionary methods and typically are approached with hand-built (non-learning) controller architectures [30–33]. Work that does involve learning typically occurs in simulation [34–37], or in relatively simple physical domains [38–42].

4   Conclusion

Can evolutionary and coevolutionary techniques be applied to real physical systems? In this paper we have presented a selection of our work each of which addresses physical evolutionary substrate in one or more dimensions. We have

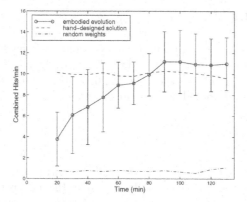

Fig. 6. Averaged runs of robots in a "light gathering" task, with various controllers.

overviewed research in handling high part-count static structures that are buildable, use of commercial CAD/CAM systems as a realistic simulated substrate, dynamic electromechanical systems with complex morphology that can be built automatically, and evolutionary techniques distributed in a physical population of robots.

Our long-term vision is that both the morphology and control programs for self-assembling robots arise directly through hardware and software coevolution: primitive active structures that crawl over each other, attach and detach, and accept temporary employment as supportive elements in "corporate" beings can accomplish a variety of tasks, if enough design intelligence is captured to allow true self-configuration rather than human re-deployment and reprogramming. When tasks cannot be solved with current parts, new elements are created through fully automatic design and rapid prototype manufacturing.

Our current research moves towards the overall goal down multiple interacting paths, where what we learn in one thrust aids the others. We envision the improvement of our hardware-based evolution structures, expanding focus from static buildable structures and unconnected groups to reconfigurable active systems governed by a central controller, and then the subsequent parallelization of the control concepts. We see a path from evolution inside CAD/CAM and buildable simulation, to rapid automatic construction of novel controlled mechanisms, from control in simulation to control in real systems, and finally from embodied evolution of individuals to the evolution of heterogeneous groups that learn by working together symbiotically. We believe such a broad program is the best way to ultimately construct complex autonomous robots who are self-organizing and self-configuring corporate assemblages of simpler automatically manufactured parts.

Acknowledgements

This research was supported in part by the National Science Foundation (NSF), the office of Naval Research (ONR), and the Defense Advanced Research Projects Agency (DARPA). Thanks to the MIT Media Lab for supplying the microcontrollers used in our embodied evolution experiments.

References

[1] Moravec, H.P.: Rise of the robots. Scientific American (1999) 124–135
[2] Morrison, J., Nguyen, T.: On-board software for the mars pathfinder microrover. In: Proceedings of the Second IAA International Conference on Low-Cost Planetary Missions, John Hopkins University Applied Physics Laboratory (1996)
[3] Nilsson, N.J.: A mobile automaton: An application of artificial intelligence techniques. In: Proceedings of the International Joint Conference on Artificial Intelligence. (1969) 509–520
[4] Brooks, R.: Intelligence without representation. Artificial Intelligence **47** (1991) 139–160
[5] Cliff, D., Noble, J.: Knowledge-based vision and simple visual machines. Philosophical Transac tions of the Royal Society of London: Series B **352** (1997) 1165–1175
[6] Lee, W., Hallam, J., Lund, H.: A hybrid gp/ga approach for co-evolving controllers and robot bodies to achieve fitness-specified tasks. In: Proceedings of IEEE 3rd International Conference on Evolutionary Computation, IEEE Press (1996) 384–389
[7] Lund, H., Hallam, J., Lee, W.: Evolving robot morphology. In: Proceedings of IEEE Fourth International Conference on Evolutionary Computation, IEEE Press (1997) 197–202
[8] Sincell, M.: Physics meets the hideous bog beast. Science **286** (1999) 398–399
[9] Dimos, D., Danforth, S., Cima, M.: Solid freeform and additive fabrication. In Floro, J.A., ed.: Growth instabilities and decomposition during heteroepitaxy, Elsevier (1999)
[10] Pollack, J.B., Blair, A.D.: Coevolution in the successful learning of backgammon strategy. Machine Learning **32** (1998) 225–240
[11] Juillé, H., Pollack, J.B.: Dynamics of co-evolutionary learning. In: Proceedings of the Fourth International Conference on Simulation of Adaptive Behavior, MIT Press (1996) 526–534
[12] Angeline, P.J., Saunders, G.M., Pollack, J.B.: An evolutionary algorithm that constructs recurrent networks. IEEE Transactions on Neural Networks **5** (1994) 54–65
[13] Sims, K.: Evolving 3d morphology and behavior by competition. In Brooks, R., Maes, P., eds.: Proceedings 4th Artificial Life Conference. MIT Press (1994)
[14] Komosinski, M., Ulatowski, S.: Framsticks: Towards a simulation of a nature-like world, creatures and evolution. In Dario Floreano, J.D.N., Mondada, F., eds.: Proceedings of 5th European Conference on Artificial Life (ECAL99). Volume 1674 of Lecture Notes in Artificial Intelligence., Springer-Verlag (1999) 261–265
[15] Floreano, D., Mondada, F.: Evolution of homing navigation in a real mobile robot. IEEE Transactions on Systems, Man, and Cybernetics (1996)

[16] Cliff, D., Harvey, I., Husbands, P.: Evolution of visual control systems for robot. In Srinivisan, M., Venkatesh, S., eds.: From Living Eyes to Seeing Machines. Oxford (1996)

[17] Lund, H.: Evolving robot control systems. In Alexander, ed.: Proceedings of 1NWGA, University of Vaasa (1995)

[18] Gallagher, J.C., Beer, R.D., Espenschield, K.S., Quinn, R.D.: Application of evolved locomotion controllers to a hexapod robot. Robotics and Autonomous Systems 19 (1996) 95–103

[19] Kawauchi, Y., Inaba, M., Fukuda, T.: Genetic evolution and self-organization of cellular robotic system. JSME Int. J. Series C. (Dynamics, Control, Robotics, Design & Manufacturing) 38 (1995) 501–509

[20] Jakobi, N.: Evolutionary robotics and the radical envelope of noise hypothesis. Adaptive Behavior 6 (1997) 131–174

[21] Mataric, M.J., Cliff, D.: Challenges in evolving controllers for physical robots. Robotics and Autonomous Systems 19 (1996) 67–83

[22] Floreano, D.: Evolutionary robotics in artificial life and behavior engineering. In Gomi, T., ed.: Evolutionary Robotics. AAI Books (1998)

[23] Funes, P., Pollack, J.B.: Computer evolution of buildable objects. In Husbands, P., Harvey, I., eds.: Fourth European Conference on Artificial Life, Cambridge, MIT Press (1997) 358–367

[24] Funes, P., Pollack, J.B.: Evolutionary body building: Adaptive physical designs for robots. Artificial Life 4 (1998) 337–357

[25] Funes, P., Pollack, J.B.: Computer evolution of buildable objects. In Bentley, P., ed.: Evolutionary Design by Computers. Morgan-Kaufmann, San Francisco (1999) 387 – 403

[26] Lipson, H., Pollack, J.B.: Evolution of machines. In: Proceedings of 6th International Conference on Artificial Intelligence in Design, Worcester MA, USA. (2000) (to appear).

[27] Lipson, H., Pollack, J.B.: Towards continuously reconfigurable robotics. In: Proceedings of IEEE International Conference on Robotics and Automation, San Francisco CA, USA. (2000) (to appear).

[28] Watson, R., Ficici, S., Pollack, J.: Embodied evolution: Embodying an evolutionary algorithm in a population of robots. In Angeline, P., Michalewicz, Z., Schoenauer, M., Yao, X., Zalzala, A., eds.: 1999 Congress on Evolutionary Computation. (1999)

[29] Ficici, S.G., Watson, R.A., Pollack, J.B.: Embodied evolution: A response to challenges in evolutionary robotics. In Wyatt, J.L., Demiris, J., eds.: Eighth European Workshop on Learning Robots. (1999) 14–22

[30] Beckers, R., Holland, O., Deneubourg, J.: From local actions to global tasks: Stigmergy and collective robotics. In Brooks, R., Maes, P., eds.: Artificial Life IV, MIT Press (1994) 181–189

[31] Balch, T., Arkin, R.: Motor schema-based formation control for multiagent robot teams. In: Proceedings of the First International Conference on Multi-Agent Systems ICMAS-95, AAAI Press (1995) 10–16

[32] Rus, D., Donald, B., Jennings, J.: Moving furniture with teams of autonomous robots. In: Proceedings of IEEE/RSJ IROS'95. (1995) 235–242

[33] Donald, B., Jennings, J., Rus, D.: Minimalism + distribution = supermodularity. Journal on Experimental and Theoretical Artificial Intelligence 9 (1997) 293–321

[34] Tan, M.: Multi-agent reinforcement learning: independent vs. cooperative agents. In: Proceedings of the Tenth International Machine Learning Conference. (1993) 330–337

[35] Littman, M.: Markov games as a framework for multi-agent reinforcement learning. In: Proceedings of the International Machine Learning Conference. (1994) 157–163

[36] Saunders, G., Pollack, J.: The evolution of communication schemes of continuous channels. In Maes, P., Mataric, M., Meyer, J.A., Pollack, J., Wilson, S., eds.: From Animals to Animats IV, MIT Press (1996) 580–589

[37] Balch, T.: Learning roles: Behavioral diversity in robot teams. In: 1997 AAAI Workshop on Multiagent Learning, AAAI (1997)

[38] Mahadevan, S., Connell, H.: Automatic programming of behavior-based robots using reinforcement learning. In: Proceedings of the Ninth National Conference on Artificial Intelligence (AAAI '91). (1991) 8–14

[39] Mataric, M.: Learning to behave socially. In Cliff, D., Husbands, P., Meyer, J.A., Wilson, S., eds.: From Animals to Animats 3, MIT Press (1994) 453–462

[40] Mataric, M.: Reward functions for accelerated learning. In Cohen, W., Hirsh, H., eds.: Proceedings of the Eleventh International Conference on Machine Learning, Morgan Kaufman (1994) 181–189

[41] Parker, L.: Task-oriented multi-robot learning in behavior-based systems. Advanced Robotics, Special Issue on Selected Papers from IROS '96 **11** (1997) 305–322

[42] Uchibe, E., Asada, M., Hosoda, K.: Cooperative behavior acquisition in multi mobile robots environment by reinforcement learning based on state vector estimation. In: Proceedings of International Conference on Robotics and Automation. (1998) 1558–1563

Biology Meets Electronics: the Path to a Bio-Inspired FPGA

Lucian Prodan[1], Gianluca Tempesti[2], Daniel Mange[2], and André Stauffer[2]

[1] Polytechnic University of Timisoara, Timisoara, Romania
E-mail: lprodan@cs.utt.ro
[2] Swiss Federal Institute of Technology (EPFL), Lausanne, Switzerland
E-mail: name.surname@epfl.ch, WWW: http://lslwww.epfl.ch

Abstract. Embryonics (embryonic electronics) is a research project that attempts to draw inspiration form the world of biology to design better digital computing machines, and notably massively parallel arrays of processors. In the course of the development of our project, we have realized that the use of programmable logic circuits (field-programmable gate arrays, or FPGAs) is, if not indispensable, at least extremely useful. This article describes some of the peculiar features of the FPGA we designed to efficiently implement our embryonic machines. More particularly, we discuss the issues of memory storage and of self-repair, critical concerns for the implementation of our bio-inspired machines.

1   Introduction

A human being is made up of some 60 trillion ($60x10^{12}$) cells. At each instant, in each of these cells, the *genome*, a ribbon of 2 billion characters, is decoded to produce the proteins necessary for the survival of the organism. The parallel execution of 60 trillion genomes in as many cells occurs ceaselessly from the conception to the death of the individual. Faults are rare and, in the majority of cases, successfully detected and repaired. This astounding degree of parallelism is the inspiration of the Embryonics (*embryonic electronics*) project [3][4][10], which tries to adapt some of the development processes of multicellular organisms to the design of novel, robust architectures for massive parallelism in silicon. The transition form carbon-based organisms to silicon-based electronics is, of course, far from immediate. Living beings exploit intricate processes, many of which remain unexplained. The Embryonics project thus focuses on two goals:

- *Similarity*: we try, where possible, to develop circuits which exploit processes similar (but obviously not identical) to those used by living organisms;
- *Effectiveness*: we wish to design systems which, while inspired by biology, remain useful and efficient from an engineer's standpoint.

As a consequence, in designing our bio-inspired computing machines, we do not try to imitate life, but rather to extract some useful ideas from some the most fundamental mechanisms of living creatures. More particularly, we are interested in the process of ontogenesis [2][12], the development of a single organism from a single cell to an adult.

J. Miller et al. (Eds.): ICES 2000, LNCS 1801, pp. 187–196, 2000.

2   Overview and Motivations

With the exception of unicellular organisms (such as bacteria), living beings share three fundamental features:

- *Multicellular organization* divides the organism into a finite number of *cells*, each realizing a unique function.
- *Cellular division* is the process whereby each cell (beginning with the first cell or *zygote*) generates one or two daughter cells. All of the genetic material (genome) of the mother cell is copied into the daughter cell(s).
- *Cellular differentiation* defines the role of each cell of the organism (neuron, muscle, intestine, etc.). This specialization of the cell is obtained through the expression of part of the genome, consisting of one or more *genes*, and depends essentially on the physical position of the cell in the organism.

Through these features, cells become "universal", as they contain the whole of the organism's genetic material (the genome), and living organisms become capable of self-repair (cicatrization) or self-replication (cloning or budding), two properties which are essentially unique to the living world.

Obviously, numerous differences exist between the world of living beings and the world of electronic circuits. To name but one, the environment that living beings interact with is continuously changing whereas the environment in which our quasi-biological development occurs is imposed by the structure of the electronic circuits, consisting of a finite (but arbitrarily large) two-dimensional surface of silicon and metal.

Taking into account these differences, we developed a quasi-biological system architecture based on four levels of organization (Fig. 1), described in detail in previous articles [3][9][10]. The particular subject of this article is the *molecular level*, a programmable circuit (FPGA) which represents the bottom layer of our system, and we will therefore discuss the other levels only insofar as they are useful for a clearer understanding of our subject matter. Notably, we will briefly describe the structure of our *cellular level* so as to justify the features we introduced in our FPGA.

3   Artificial Cells

As shown in Fig. 2, our artificial organisms are divided into a finite number of cells. Each cell is a simple processor (a binary decision machine) which realizes a unique function within the organism, defined by a set of instructions (program), which we will call the *gene* of the cell. The functionality of the organism is therefore obtained by the parallel operation of all the cells.

Each cell stores a copy of the genes of all the cells of the organism (which, together, represent the operative part of our artificial genome, or, in short, the *operative genome*), and determines which gene to execute depending on its position (X and Y coordinates) within the organism, implementing *cellular differentiation*. For example, in Fig. 2 each cell of a 6-cell organism realizes one of the six possible genes (A to F), but stores a copy of all the genes.

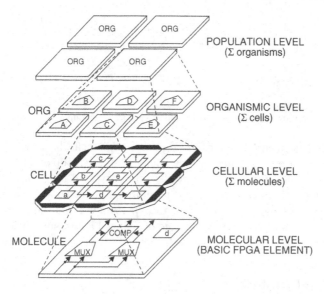

Fig. 1. The four levels of organization of Embryonic systems.

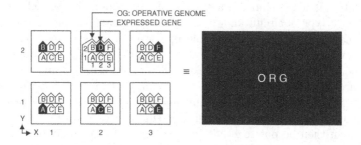

Fig. 2. The multi-cellular organization of our artificial organisms.

3.1 Cyclic vs. addressable memory implementation

One of the issues raised by our approach is the implementation of memory: since each cell must contain the entire genome, an important percentage of its silicon surface will consist of read-only program memory. Conventional addressable memory, while of course capable of the task, requires relatively complex addressing and decoding logic, contrary to our requirement that the cells to be as simple as possible. To store the operative genome in our artificial cell, we therefore studied alternative ways to implement the required memory.

In living cells, the genetic information is processed *sequentially*, suggesting a different type of memory, which we will call *cyclic memory*. Cyclic memory does not require any addressing. Instead, it consists of a simple storage structure that continuously shifts its data in a closed circle (Fig. 3). Data is accessed sequentially, much as the ribosome processes the genome inside a living cell [1].

Fig. 3. Comparison between addressable (left) and cyclic (right) memories.

Cyclic memory, while simpler to design and implement, has several disadvantages when compared to its addressable analog. Notably, loop execution becomes very onerous, as the entire memory needs to be traversed for each iteration of a loop, however small. Such a memory is therefore not adapted for general-purpose applications (particularly for large programs). However, for the particular case of Embryonics, we feel that the advantage of reduced addressing logic overcomes the drawbacks, and thus opted for a cyclic memory to store our genome program.

4   Artificial Molecules

The lowest, most basic level of organization of our system is the molecular level, implemented as a two-dimensional array of programmable logic elements (the molecules), a type of circuit known as a *field-programmable gate array* (FPGA).

Our reasons for introducing a molecular level in our systems is explained in detail elsewhere [3][9][10]. Essentially, we require a substrate of programmable logic to be able to adapt the size and structure of our cells to a given application. FPGAs are an obvious answer to this problem: they provide us with a uniform surface of programmable elements (our *molecules*) which can be assigned a function at runtime via a software configuration (in our case, the MOLCODE). These elements can then be put together through a set of programmable connections to realize virtually any digital circuit, and notably our artificial cells (Fig. 4).

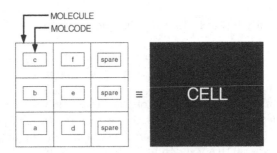

Fig. 4. Multi-molecular structure of our artificial cells.

Our new FPGA, called *MuxTree* [3][9], implements all of the features required of our artificial molecules to efficiently implement our bio-inspired computing machines.

4.1   Architecture

The functional unit FU of our artificial molecule (Fig. 5) is based on a multiplexer (hence the name of MuxTree for *multiplexer tree*) coupled with a D-type flip-flop to implement sequential systems. The functional unit is duplicated to perform self-test (briefly introduced below). The molecule also features two sets of connections: a fixed, directional network for communication between neighbors, and a programmable, non-directional network for long-distance communication (routed through the switch block SB). The bits required to assign the molecule's functionality are stored in the 20-bit wide shift register CREG.

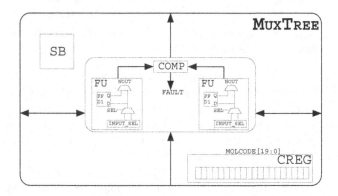

Fig. 5. Architecture of our artificial molecule.

The configuration of the array is effected by chaining together the shift registers of the molecules (Fig. 6). A small cellular automaton [3][11] (the *space divider*), placed *between* the molecules, channels the configuration bitstream to the correct molecules (avoiding, if necessary, faulty molecules). The space divider also lets a single bitstream be replicated as many times as required, allowing the automatic realization of arrays of identical processing elements from the description of a single such element (cellular division). Finally, it lets the user decide which (if any) of the columns of molecules within the array will be *spare*, that is, held in reserve to allow faults to be repaired (see below).

4.2   The memory mode

As we have seen, the genome memory, implemented as a cyclic memory, is a fundamental part of our cells. With our FPGA, it is of course possible to realize this memory using the storage elements of each molecule (the D flip-flops). However, storing a single bit of information in each molecule is very inefficient.

We thus decided to exploit the configuration register CREG to implement our genome memory. This register has two remarkable advantages from this point of view: it is a shift register, an ideal component for the realization of cyclic memories, and it already possesses a set connections (those used for the propagation of the configuration bitstream) to chain together multiple registers.

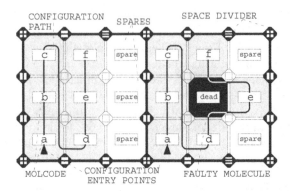

Fig. 6. Configuration of the FPGA as two copies of an identical cell, with one spare column per cell and one faulty molecule.

In our FPGA, each molecule can thus have two modes of operation: the *logic mode*, where the molecules are programmable logic elements realizing combinational and sequential circuits, and the *memory mode*, where the molecules are storage elements for a cyclic memory. In memory mode, each molecule can store either 16 or 8 bits of information (in the latter case, the long-distance connection switch block SB remains operational).

Each molecule can store 16 or 8 1-bit-wide words, and multiple molecules can be chained together (using the existing connections) to realize deeper and/or wider memories (Fig. 7). These memory blocks can then be directly exploited by our cellular processors to hold the genome program, without the overhead associated with the control and addressing of conventional memories.

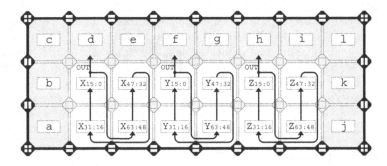

Fig. 7. A cyclic memory of 64 3-bit wide words set inside a cell.

5   Self-Repair

Biological entities are continuously put under environmental stress. Wounds and illnesses resulting from such stress often cause incapacitating physical alterations. Fortunately, living beings are capable of successfully fighting the great majority of such wounds and illnesses, showing a remarkable robustness through a process that we call *healing*.

To endow our artificial organisms with similar features, we provide a two-level mechanism for self-repair, involving both the cellular and the molecular level. What follows is a description of healing at the cellular and molecular level and of the way the two levels cooperate to produce a higher level of robustness than would be allowed by a single level.

5.1 Self-Repair at the cellular level

The redundant storage of the entire genome in every cell is obviously expensive in terms of additional memory. However, it has the important advantage of making the cell *universal*, that is, potentially capable of executing any one of the functions required by the organism. This property, coupled with the coordinate-based gene selection, is a huge advantage for implementing *self-repair* (i.e., healing), the process behind the formidable robustness of living systems.

In fact, since our cells are universal, and since their gene is determined by the cell's coordinates, the system can survive the "death" of any one cell simply by re-computing the cells' coordinates within the array, provided of course that "spare" cells (i.e., cells which are not necessary for the organism, but are held in reserve during normal operation) are available (Fig. 8).

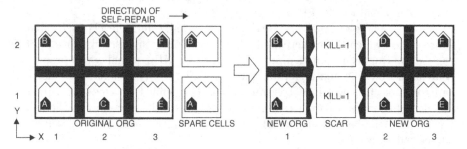

Fig. 8. Self-repair of the cellular level through coordinate recomputation.

Self-repair at the cellular level thus consists simply of deactivating the column containing the faulty cell: all the cells in the column "disappear" from the array, that is, become transparent with respect to all horizontal signals. Since the coordinates are computed locally from the neighbors' coordinates, any disappearance forces all the cells to right of the dead column to recalculate their coordinates, completing the reconfiguration of the array [6][10].

5.2 Self-Repair at the molecular level

In order to avoid the stiff penalty inherent in killing a column of processors for every fault in the array, we introduced a certain degree of fault tolerance at the molecular level. Self-repair in an FPGA implies two separate processes: self-test and reconfiguration.

Of these, self-test is undoubtedly the costliest, requiring a complex hybrid solution mixing duplication and fixed-pattern testing, detailed elsewhere [9].

On the other hand, the homogeneous architecture of our FPGA simplifies reconfiguration to a considerable extent [5][7]. Since all molecules are identical, and the connection network is homogeneously distributed throughout the array, reconfiguration becomes a simple question of shifting the configuration of the faulty molecule to its right (similarly to what happens during configuration, as shown in Fig. 6) and redirecting the array's connections. This procedure can be accomplished quite easily, assuming that a set of spare molecules are available. The determination of these spare molecules is in fact one of the most powerful features of our system, since we can exploit the space divider to dynamically allocate some columns as spares. The position and frequency of spares can then be determined at configuration time, and the fault tolerance of our FPGA becomes programmable (and can thus be adapted to the circumstances and the operating conditions).

5.3   Two-level self-repair: the KILL signal

Even if the robustness of the self-repair mechanism at the molecular level is programmable, there are limits to the faults which can be repaired at this level. Notably, if a fault is large enough to affect multiple adjacent molecules on the same row or if it occurs in a non-repairable part of the molecule, no amount of redundancy will let the FPGA repair itself.

As we have seen, however, there exists a second, cellular level of self-repair in our system, which can be activated whenever the molecular self-repair fails. This activation can occur quite simply by generating a KILL signal which propagates outwards from the non-repairable molecule (Fig. 9). This signal will "destroy" all the molecules within a column of cells (as defined by the space divider), rendering them transparent to horizontal signals (in the same way as the unused spare molecules are transparent), thus activating the cellular-level self-repair described above.

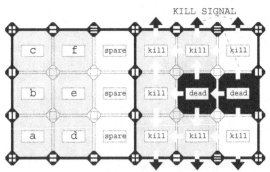

Fig. 9. Faults in adjacent molecules activate the KILL signal.

The molecular and the cellular levels of our system thus cooperate to assure a high degree of robustness to the system: when the molecular level cannot handle the self-repair by itself, it activates the cellular level so that the organism can continue to survive.

5.4   The UNKILL mechanism

In digital electronic systems, the majority of hardware faults occurring in the silicon substrate are in fact *transient*, that is, disappear after a short span of time. This observation is an important issue when designing self-repairing hardware: the parts of the circuit which have been "killed" because of the detection of a fault could potentially come back to "life" after a brief delay.

Detecting the disappearance of a fault and handling the "unkilling", however, usually requires a relatively complex circuit. This complexity prevents us from implementing this feature at the molecular level (the molecules being very small and simple components). At the cellular level, on the other hand, it is not only possible, but also quite simple to implement this feature.

We have seen that self-repair at the cellular level consists of "destroying" (i.e., resetting) the configuration of all the molecules within a column of cells, with the effect of making the column "invisible" to the array. As it has been reset, however, nothing prevents us from sending once more the configuration bitstream to the deactivated molecules (Fig. 10): if some or all the faults have vanished, then the configuration will restore the functionality of the cells, and the reappearance of the column within the array will engender a new re-computation of the array's coordinates (in the direction opposite to self-repair) and restore the entire array to its original size and functionality.

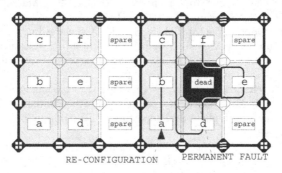

Fig. 10. While the left-hand side cell operates, the right-hand side cell is "unkilled".

6   Conclusions and Future Directions

The Embryonics project has been progressing for many years. Throughout this time, we have been accumulating considerable experience in trying to adapt biological concepts to the world of electronics. Of course, Embryonics represents only one of the many possible approaches to bio-inspiration. However, we feel that other projects drawing inspiration from the ontogenetic development of living beings for the design of computer hardware will face many of the same problems, and are likely to find solutions not too dissimilar from our own.

In particular, to implement machines inspired by the ontogenetic development of living beings, the use of programmable circuits is (at least in the short term) likely to be a requirement, imposed by the need to vary the cellular structure as a function of the application. Moreover, such a circuit will need to support some form of self-replication, to handle the massive amount of cells in a complex organism, to be able to tolerate faults (both permanent and transient), because of the sheer amount of silicon required, and to efficiently store the important amount of memory required by a genome-based approach, without which an ontogenetic machine would be impossible.

Of course, in the long run, technological advances (for example, the development of nanotechnologies) will probably render many of our specific mechanisms obsolete. But even then, our overall approach to the problem of designing ontogenetic hardware might well remain valid, if not even assume greater importance (self-replication [8], for example, is one of the key issues in nanotechnology).

Acknowledgements

This work was supported in part by the Swiss National Foundation under grant 21-54113.98, by the Consorzio Ferrara Richerche, Università di Ferrara, Ferrara, Italy, and by the Leenaards Foundation, Lausanne, Switzerland.

References

1. Barbieri, M.: The Organic Codes: The Basic Mechanism of Macroevolution. Rivista di Biologia / Biology Forum **91** (1998) 481–514.
2. Gilbert, S. F.: Developmental Biology. Sinauer Associates Inc., MA, 3rd ed. (1991).
3. Mange, D., Tomassini, M., eds.: Bio-inspired Computing Machines: Towards Novel Computational Architectures. Presses Polytechniques et Universitaires Romandes, Lausanne, Switzerland (1998).
4. Mange, D., Sanchez, E., Stauffer, A., Tempesti, G., Marchal, P., Piguet, C.: Embryonics: A New Methodology for Designing Field-Programmable Gate Arrays with Self-Repair and Self-Replicating Properties. IEEE Transactions on VLSI Systems *6(3)* (1998) 387–399.
5. Negrini, R., Sami, M.G., Stefanelli, R.: Fault Tolerance Through Reconfiguration in VLSI and WSI Arrays. The MIT Press, Cambridge, MA (1989).
6. Ortega, C., Tyrrell, A.: Reliability Analysis in Self-Repairing Embryonic Systems. Proc. 1st NASA/DoD Workshop on Evolvable Hardware, Pasadena, CA (1999) 120–128.
7. Shibayama, A., Igura, H., Mizuno, M., Yamashina, M. An Autonomous Reconfigurable Cell Array for Fault-Tolerant LSIs. Proc. 44th IEEE International Solid-State Circuits Conference, San Francisco, California (1997) 230–231,462.
8. Sipper, M. Fifty Years of Research on Self-Replication: an Overview Artificial Life *4(3)* (1998) 237–257.
9. Tempesti, G.: A Self-Repairing Multiplexer-Based FPGA Inspired by Biological Processes Ph.D. Thesis No. 1827, EPFL, Lausanne (1998).
10. Tempesti, G., Mange, D., Stauffer, A.: Self-Replicating and Self-Repairing Multicellular Automata. Artificial Life *4(3)* (1998) 259–282.
11. Wolfram, S.: Theory and Applications of Cellular Automata. World Scientific, Singapore (1986).
12. Wolpert, L.: The Triumph of the Embryo. Oxford University Press, NY (1991).

The Design and Implementation of Custom Architectures for Evolvable Hardware Using Off-the-Shelf Programmable Devices

Craig Slorach and Ken Sharman

Department of Electronics and Electrical Engineering, University of Glasgow
Glasgow, Scotland, G12 8LT, United Kingdom
{c.slorach, k.sharman}@elec.gla.ac.uk
http://www.elec.gla.ac.uk/~craigs/ehw.html

Abstract. In this paper we present the design and implementation of architectures suitable for evolving digital circuits and show how these architectures can be instantiated on 'off-the-shelf' commercial programmable logic devices such as Field Programmble Gate Arrays (FPGA's). We discuss architecture details and the design flow from the initial cell specification through to targeting specific commercial devices. We also present an example of such an implementation, where a simple fine-grained programmable logic array is designed and targeted to a number of popular commercial FPGA's. A critical discussion of this approach is also given.

1 Introduction

In order for intrinsic [1] Evolvable Hardware (EHW) to be viable, there is a key need for supporting programmable logic devices to be freely available and in plentiful supply. The Xilinx XC6200 series [2] was the first commercially available programmable device that supported the requirements for evolvable hardware and has been used extensively in research [3] [4]. The key feature that allowed the XC6200 to be suitable for EHW research is that no instantiated circuit on the device could lead to it being damaged. However, this device is no longer available and there are no current devices on the market that readily support EHW due to the specific device requirements it imposes (discussed further in Section 2).

There have already been solutions developed which have counteracted this problem. At the most extreme, various researchers [5] have developed custom LSI solutions for EHW. However, these solutions are expensive to develop and hence for custom research unviable. Another option is to use a software approach to work with other programmable devices, for example the JBits package [6], however this approach tends to involve a software overhead and also restricts the possibility of solutions by imposing constraints on the possible space of circuits that could be evolved.

One possible solution to the above problem is to use commercial programmable devices (such as Field Programmable Gate Arrays- FPGA's) for EHW by writing

J. Miller et al. (Eds.): ICES 2000, LNCS 1801, pp. 197–207, 2000.

a programmable structure directly onto a conventional device which can then be used to evolve logic circuits directly on in an intrinsic manner. The use of such a 'system on a system' approach is already employed extensively in the semiconductor design industry where large prototyping systems are available [7] which contain commercial FPGA's which are used to evaluate large digital circuits before production. We propose the use of such an approach in this paper.

The paper is structured as follows: In Section 2 we discuss the various architecture issues relating to EHW, describing the architecture requirements for EHW and also presenting a generic architecture. Section 3 describes the key issues for employing commercial FPGA's for implementing such structures. Section 4 describes the design flow for taking a design through to any commercial device, Section 5 describes an example implementation of a fine-grained logic device targeted to various logic families. Section 6 then discusses other issues relating to the work, followed by a conclusion in Section 7.

2   Evolvable Hardware Device Architecture

2.1   EHW Device Goals/ Requirements

In considering an architecture suitable for EHW, it is important that key requirements for EHW work must be met, these are discussed extensively in [8].

First and foremost, the device should not be damaged by incorrect programming bits being sent to the device. All current commercial FPGA's do not support this requirement- it is possible to send programming bitstreams which can damage the device. A software approach may be possible to overcome this limitation as discussed in [9] though it is more desirable for the device to handle this directly (i.e. all bitstreams create 'valid' circuit configurations which cannot damage the device).

To aid *genotype* to *phenotype* mapping, each cell in the device should be addressed individually so that it is easy to map the gene to circuit configurations (i.e. each cell is represented by a small subrange of the configuration bitstream). Spatial independence is also important here, i.e. every 'independent' cell has similar gene-phene mappings. Given that speed is of paramount importance, only cells that are required for the target circuit should need to be programmed during the evolution phase- i.e. there is no need to generate an entire device bitstream if only a small portion of the device is being used. Configuration should also be able to be performed 'in-circuit' without the need for complex interface hardware, ideally to generate the configuration directly from the evolutionary algorithm in cell address/ configuration data pairs.

A degree of dynamic reconfiguration would be useful if parallel implementations of the phenotype were to be employed- this would allow the evaluation of individuals whilst others are being updated.

2.2   Generic Device Architecture

A generic device architecture is depicted in Fig. 1. This shows the array of *cells* that are present in the device along with how they are interconnected. The

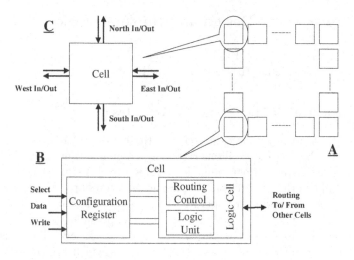

Fig. 1. Generic programmable device architecture with local routing and configuration register. 'A' shows the array of cells present in the device. 'B' shows the internals of the cell. 'C' shows local routing between cells.

maximum number of cells that can be implemented on an FPGA is a function of the degree of logic required to implement the cell and the size of the target FPGA. Each cell comprises a *logic cell* and *configuration register*.

The logic cell implements a range of programmable logic functions. It also implements programmable routing. A sub-block of the logic cell is the *logic unit* which contains the circuit which allows different logic systems to be implemented. For example the logic unit could contain a Universal Logic Module (ULM) [10] with controllable inputs or could contain significantly larger amounts of programmable logic if so desired (for example, implement a function with a larger number of inputs or a more diverse range of possible logic functions). The routing control section controls the routing to and from the logic cell. In Fig. 1 only local routing is shown, where the data inputs to the logic unit can come from any of the nearest (N, S, E, W) neighbouring cells and the direction outputs from the cell can be any of the other direction inputs or the output from the logic unit.

Each cell has its own local configuration register. The configuration register controls the operation of the logic cell (i.e. routing and logic function). The size of the configuration register is dependent on the degree of control logic required for routing and logic function control. Each configuration register in the design is selected by some function of a subset of the address bus. A global data bus is present around the device and is connected to each configuration register for programming purposes. Writing in new configuration data is achieved by a global control signal which is set when writing new configuration data for a cell.

3   Key Issues for Design and Implementation

In order to use the target FPGA resources efficiently, there are several issues that must be considered, these are addressed below:

3.1   Efficient use of Logic and Routing Resources

Given that the logic resources present on the device are in short supply, it is clear that the best use of available resources (combinational logic and flip-flops) are made use of. Most programmable architectures generally have logic blocks which contain both combinatorial logic and also flip-flops. To ensure that the device has the opportunity to operate as fast as possible, routing should also be kept to a minimum, with routing kept as local as possible.

For configuration of each of the programmable cells, several options are possible. One possibility is to instance RAM macros which are present for most commercial FPGA's, however such an approach leads to a device specific architecture. Additionally, the RAM present in macros is of the Nx1 style (single bit) as opposed to the 1xN style (multiple bit) that would be required for persistent configuration control. As a consequence, we conclude that the best approach to use Flip-Flops to act as configuration registers. Furthermore, the configuration registers should be localised to each cell in the design to make use of Flip-Flops which are available close to the logic used to implement the logic cell.

3.2   Startup Configuration

It is also important to consider the configuration that any EHW device will have on startup. Specifically, it is important that no asynchronous feedback circuits be created. This can be easily achieved by ensuring that at startup each cell in the device is configured to feed-forward signals and not route them back into previous cells.

3.3   Use of Hierarchy

In order to maximise flexibility and reuse, the design should exploit hierarchy. Ideally, the design of a single configurable cell should be undertaken as a first step. Multiple instances of this cell should then be used to form groups of cells (for example, a 4 cell array). These groups of cells can then in turn be used to form larger configurable arrays, for example a 16 cell array. This approach can then be extended until the desired number of cells are achieved.

4   Design Flow

Fig. 2 describes the design flow from taking the design architecture to a working EHW programmable device.

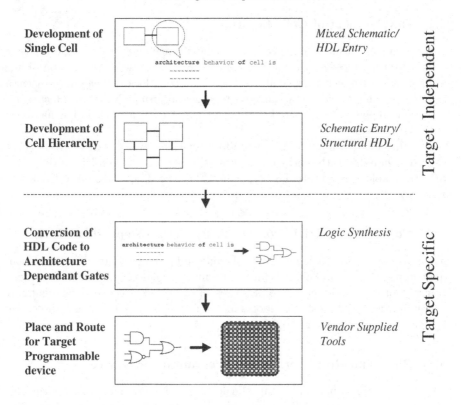

Fig. 2. Design flow for implementation of custom architectures.

4.1   Development of a Single Cell

The first stage of the implementation is to design a single cell for the device. There are various design issues which should be to considered at this stage. First and foremost is the design of the logic unit, specifically how much logic has to be available in the logic unit- this will vary from application to application. At the simplest level, the logic unit could provide a 2 input programmable logic function, or could be extended to a more complex logic unit (e.g. 4 input function). Routing between cells must also be considered, specifically what routing is to be available and how it is to be implemented.

After these design considerations have been specified, the next stage is to create a single cell. This is generally achieved by mixed Hardware Description Language (HDL) and schematic entry. For example, the logic cell could be composed of a schematic containing a number of symbols of primitive elements and these could be then described using an HDL such as VHDL or Verilog.

The output from this stage is a complete description of the cell.

4.2   Development of Cell Hierarchy

After a single cell has been specified, the next stage is to create a hierarchy of cells. This is required as it is inconvenient to create a single schematic containing a large number of cells. Typically, a 2 cell by 2 cell block is created and then a number of these blocks are connected together to form a NxM cell block. This process is repeated the desired number of times until a final design has been realised.

At this stage, a final design in the form of HDL source code is available. It should be noted that the entire design upto this point is independent of the target programmable architecture, i.e. a specific FPGA need not have been decided at this step.

4.3   Conversion of HDL Code to Architecture Dependent Gates

The next stage is to synthesize the HDL code and produce an architecture dependent implementation. This is typically performed using a commercial synthesis tool (Synopsys, Ambit, etc.) and produces an FPGA architecture dependent gate representation. At this stage, reports may be generated and viewed to ascertain the gate requirements for the design.

4.4   Place and Route for Target Programmable Device

The final stage is to place and route the design for the programmable device. This step involves the use of FPGA vendor supplied place and route tools. The output from this step is a configuration bitstream for the EHW device. Reports are also available at this stage providing details of the device utilization and also the target speed of the device (specifically, how fast the device can be programmed by the host evolutionary algorithm and also the specifications for the logic cells in terms of propagation delay).

5   An Example Implementation

In order to illustrate the approach discussed above, we provide an example implementation in this section and show how it can be targeted to a number of commercially available programmable logic devices. The design is a simple fine-grained architecture similar to the Xilinx XC6200 series [2].

5.1   Logic Unit

A suitable logic unit for a fine-grained implementation is given in Fig. 3. The logic unit is based around a Universal Logic Module (ULM) [10], which exploits the fact that a 2 input multiplexer can implement many different logic functions depending on the choice of inputs (in this case from any of the given directions-N, S, E, W). Each of the inputs to the ULM are controlled by 4 input multiplexers

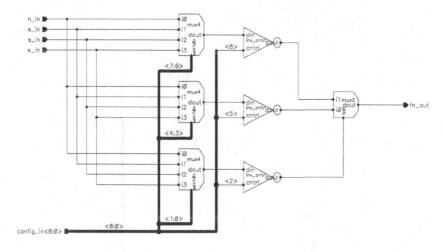

Fig. 3. Logic Unit for Example Implementation.

and there are also inverters to apply either the selected direction input or its complement.

The multiplexers and inverter control symbols are described in elementary VHDL.

5.2   Routing/ The Complete Logic Cell

The complete logic cell showing both the logic unit and routing control is shown in Fig. 4. Routing is achieved via the use of 4-1 multiplexers which control the direction of outputs from the cell.

5.3   Configuration Issues

Given that there is an abundance of freely available flip-flops in most FPGA's (as discussed in Sect. 3.1) the configuration registers can be implemented by Flip-Flops, the output from these flip-flops is then fed into the configuration inputs (i.e. the control inputs of the multiplexers). The use of registers for configuration creates the need for a configuration clock to be present on the EHW device.

In order to ensure that the device starts up in a non-critical configuration, the cell outputs are configured in a 'feed-forward' manner, this ensures that no feedback circuits are implemented at startup. Initialisation is performed by a global reset signal which is connected to all the configuration registers.

5.4   Complete Cell Implementation

The final cell implementation is given in Fig. 5. This shows the logic cell and the configuration register used to control the logic configuration and routing. Each cell requires a total of 17 configuration bits.

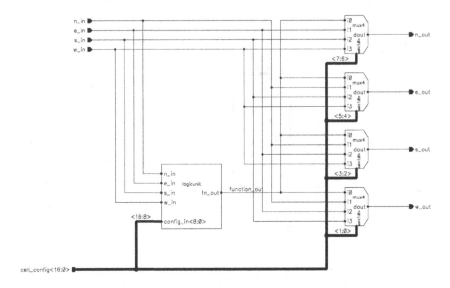

Fig. 4. Logic Cell for Example Implementation.

5.5   The Top Level Design

A top level design containing 16 complete cells was then designed in schematic form. This was undertaken in a hierarchical manner, with a 2x2 cell block first created and then 4 of these used to form a 16 cell device, etc.

5.6   Results

In order to validate the design, we chose a simple design of 16 cells in a 4x4 structure. We synthesized the top level design using Synopsys FPGA Express, a standard commercial synthesis tool. Results are presented for a number of commercial FPGA's in Table 1. [1] These results show that the 4x4 cell design (based on the cells described in Section 5.4) occupies 57% of the resources in a Xilinx 4010E (a popular commercial FPGA). Furthermore, we conclude that a maximum of 28 configurable cells will fit into this specific device.

Table 1. Results from 4x4 Cell Implementation.

Device	Gate Count	CLBs	% Occupation	Predicted Max. Cells
Xilinx XC4010E	7-20K	950	57	28
Altera EPF6010A	5-10K	880	66	24

---

[1] Gate count taken from manufacturer datasheet, for guidance only.

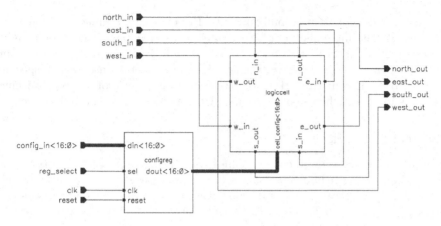

Fig. 5. Final configurable cell.

6   Discussion and Future Work

6.1   Scalability Issues

It is clear from Table 1 that only a small number of cells will fit onto the devices quoted. However, it should be stressed that these devices (used to demonstrate the concept) are relatively small in current commercial terms. Consequently, it will be possible to put significantly more cells on larger devices which are readily available.

Given that the main routing of the cell is highly localised, we suggest that the logic block requirements of implementation will scale linearly with the number of cells required in the design. The size of each cell is dependent on the complexity of the programmable logic function required.

6.2   Timing Issues

There are two key timing issues which affect the suitability of our suggested approach- re-configuration speed and propagation delays.

Re-configuration speed is influenced partially by the maximum operating speed of the Flip-Flops (generally high) and also the delays incurred in routing the configuration data (i.e. to the local configuration registers for each cell) which can be controlled during the place and route stage of the design.

After reconfiguration, the maximum operating speed (for example, when evaluating a circuit for fitness) is dictated by propagation delays in the target FPGA and will also vary with the number of cells used to evolve circuits.

6.3   Implementation Overhead

Clearly, the approach we discuss above is by no means the most efficient use of programmable logic resources which already pose a significant burden in terms of transistor count for their reconfigurability. However, programmable devices are increasingly offering higher and higher gate counts (1M gates is now common in devices such as the Xilinx Virtex [11]), so this will become less of an issue from an implementation perspective. It may also be possible to use vendor specific macros to implement the EHW structure and also the use of a vendor-specific design is likely to achieve better results at the expense of vendor independence.

6.4   Future Work

Future work will consider the use of high-density devices such as the Xilinx Virtex [11]. Other logic unit implementations will also be considered. Additionally, the use of programmable processor IP cores will also be explored to create System on a Chip (SoC) single device EHW solutions containing a general purpose processor and programmable logic array.

7   Conclusion

In conclusion, we have demonstrated that although no current commercial programmable logic devices fully support evolvable hardware, custom-built programmable devices can be suitably realised on conventional FPGA's by implementing a user-defined structure using conventional CAD software. Specifically, we have demonstrated that a fine-grained programmable architecture suitable for evolving digital circuits on can be implemented on commonly available FPGA's.

Acknowledgements

This research is funded through an EPSRC studentship. We wish to thank Xilinx and Altera for provision of place and route software for their programmable devices. This work has also made use of hardware resources provided by a SHEFC Research Development Grant.

References

1. H. de Garis, Growing an Artificial Brain with a Million Neural Net Modules Inside a Trillion Cell Cellular Automation Machine, In Proc. 4th Symp. on Micro Machine and Human Science, pages 211-214, 1993.
2. Xilinx Inc., XC6200 Field Programmable Gate Arrays Datasheet V1.10, San Jose, CA, 1997.
3. T.C. Fogarty, J.F. Miller, P. Thompson, Evolving Digital Logic Circuits on Xilinx 6000 Family FPGA's, In P.K. Chawdry, R. Roy, R.K. Pant, eds., Soft Computing in Engineering Design and Manufacturing, pages 299-305, Springer-Verlag, London, 1998.
4. A. Thompson, An Evolved Circuit, Intrinsic in Silicon, Entwined with Physics, In Tetsuya Higuchi, Masaya Iwata, Weixin Liu, eds., Evolvable Systems: From Biology to Hardware, First International Conference, ICES 96, pages 390-405, Springer-Verlag, Berlin, October 1996. Lectures Notes in Computer Science 1259.

5. I. Kajiani, *et al*, A Gate-Level EHW Chip: Implementing GA Operations and Reconfigurable Hardware on a Single LSI, In Moshe Sipper, Daniel Mange, Andres Perez-Uribe ,eds., Evolvable Systems: From Biology to Hardware, Second International Conference ICES 98, pages 1-12, Springer-Verlag, Berlin, September 1998. Lecture Notes in Computer Science 1478.

6. S.A. Guccione, D. Levi, XBI: A Java-Based Interface to FPGA Hardware, In John Schewel, editor, Configurable Computing: Technology and Applications, Proc. SPIE 3526, pages 97-102, Bellingham, WA, November 1998. SPIE – The International Society for Optical Engineering.

7. Aptix Inc., System Explorer Datasheet, http://www.aptix.com/nova/products.

8. H. de Garis, Evolvable Hardware: Genetic Programming of a Darwin Machine, In R.F. Albrecht, C.R. Reeves, N.C. Steele, eds., Artificial Neural Nets and Genetic Algorithms, Springer-Verlag, NY, 1993.

9. D. Levi, S.A. Guccione, GeneticFPGA: A Java-Based Tool for Evolving Stable Circuits, In John Schewel, editor, Reconfigurable Technology: FPGAs for Computing and Applications, Proc. SPIE 3844, pages 114-123, Bellingham, WA, September 1999. SPIE – The International Society for Optical Engineering.

10. C.E. Shannon, The Synthesis of Two-Terminal Switching Circuits, Bell System Technical Journal, Vol. 28, pages 59-98, January 1949.

11. Xilinx Inc., Virtex 2.5V Field Programmable Gate Arrays Datasheet V1.7, San Jose, CA, 1999.

Mixtrinsic Evolution

Adrian Stoica, Ricardo Zebulum and Didier Keymeulen

Center for Integrated Space Microsystems
Jet Propulsion Laboratory
California Institute of Technology
Pasadena CA 91109, USA

Abstract. Evolvable hardware (EHW) refers to automated synthesis/optimization of HW (e.g. electronic circuits) using evolutionary algorithms. *Extrinsic EHW* refers to evolution using software (SW) simulations of HW models, while *intrinsic EHW* refers to evolution with HW in the loop, evaluating directly the behavior/response of HW. For several reasons (including mismatches between models and physical HW, limitations of the simulator and testing system, etc.) circuits evolved in SW may not perform the same way when implemented in HW, and vice-versa. This *portability problem* limits the applicability of SW evolved solutions, and on the other hand, prevents the analysis (in SW) of solutions evolved in HW. This paper introduces a third approach to EHW called *mixtrinsic EHW (MEHW)*. In MEHW evolution takes place with hybrid populations in which some individuals are evaluated intrinsically and some extrinsically, within the same generation or in consecutive ones. A set of experiments using a Field Programmable Transistor Array (FPTA) architecture is presented to illustrate the portability problem, and to demonstrate the efficiency of mixtrinsic EHW in solving this problem.

1 Introduction

Evolvable HW (EHW) refers to automated synthesis/optimization of HW (e.g. electronic circuits) using evolutionary algorithms. Previous reports remark that solutions obtained by evolutionary design suffer from what it will be referred in this paper as *the portability problem*. For example, it was observed that some circuits obtained through evolutionary design on one HW platform had a different behavior when tested on a second platform, although the two were of similar type/construction [1]. Furthermore, a similar situation was encountered when attempting to port to HW the result of a solution evolved in SW, and vice-versa [2].

Evolution based on simulations of HW models is referred to as *extrinsic evolution (EE)*, while evolution with the HW in the loop (evaluating directly the behavior/response of HW) is referred to as *intrinsic evolution (IE)*. Successful accounts of intrinsic EHW (IEHW) and extrinsic EHW (EEHW) are reported in the literature [3-5]. However the portability between the SW and HW implementations of the solutions has been difficult. Researchers who evolved extrinsically often lacked suitable programmable devices to test their solutions (particularly for the evolution of analog circuits). In turn, those evolving intrinsically had limited access to the SW models of their HW platform, which often was proprietary information. Only

J. Miller et al. (Eds.): ICES 2000, LNCS 1801, pp. 208–217, 2000.

recently, and mainly through evolving on in-house built devices/test-boards, it became apparent that mismatches may exist, and the solutions evolved in SW may not hold in HW and vice-versa [2]. To solve this portability problem, a third approach, called *mixtrinsic EHW,* is proposed in this paper. Mixtrinsic EHW encompasses a family of techniques that combine the intrinsic and extrinsic modes in a variety of ways. The most straightforward alternative is the use of a mixed population of both SW models and reconfigurable HW.

This paper illustrates the portability problem with several examples, and presents a solution to this problem using mixtrinsic evolution (ME). The paper is organized as follows: Section 2 discusses characteristic aspects of extrinsic and intrinsic evolution. Section 3 focuses on the portability problem. Section 4 introduces a novel approach to EHW called mixtrinsic EHW (MEHW). Section 5 reviews the main characteristics of a Field Programmable Transistor Array (FPTA) used in the following sections as an evolutionary platform to demonstrate MEHW. Section 6 illustrates the portability problem and demonstrates how MEHW can solve it by exploiting common characteristics of the SW and HW. It also discusses the opposite effect, i.e. how MEHW can be used to emphasize differences between SW and HW and possibly capitalize on characteristics of physical HW. Section 7 presents the conclusions.

2   Extrinsic and Intrinsic Evolution

Two approaches to EHW have been proposed. The first uses *extrinsic* evolution, the candidate solutions are evaluated as SW models (of HW) and evaluations are done using a simulator. EEHW is schematically illustrated in Fig. 1. The population is homogeneous, and consists of SW models (e.g. SPICE netlists) that describe an electronic circuit to a certain degree of accuracy. The second approach to EHW uses intrinsic evolution, where the candidate solutions are in the form of physical HW configurations on programmable devices/architectures, which are evaluated using some test/evaluation equipment. IEHW is illustrated in Fig. 2. IEHW is more sensitive as the candidate solutions can a) damage the chip in some overstressing conditions, and b) be affected by previously configured/tested candidates. While in EEHW, individual candidates have no influence on each other, in IEHW they do, because each candidate performs on the same "stage" as its predecessors, which may have left an imprint on the "stage". For example, the charge accumulated during the evaluation of the performance of one individual affects the behavior of the next.

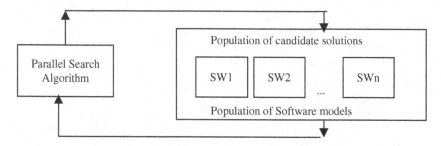

Fig. 1. Extrinsic EHW: evaluations of software solutions

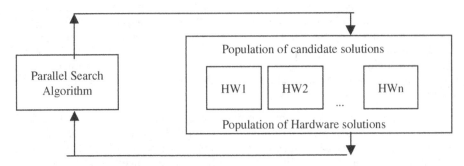

Fig. 2. Intrinsic EHW: evaluations of hardware solutions

3   The Portability Problem

Early experiments in EHW made it apparent that the solutions obtained by evolutionary design may suffer from a portability problem. For example, it was observed that some circuits obtained through evolutionary design on one HW platform had a different behavior when tested on a second platform, although the two were of similar type/construction. Thus, a circuit evolved on a corner of an FPGA did not reproduce the same behavior when it was implemented on a different part of the same FPGA [1]. Another situation is related to porting to HW a circuit evolved in SW (or vice-versa validating in SW a solution evolved in HW) as reported in [2]. Some of the circuits resulting as solutions from extrinsic evolution do not produce the same correct response when implemented/ported into HW. Vice-versa, many of the circuit topologies resulting from intrinsic evolution do not produce a good response (as obtained in the real HW) when they are simulated in SW.

One reason behind the portability problem is that, in each case, evolution finds the easy way out, optimizing for whichever raw material is given. The portability problem between two HW platforms is strongly related to differences in a set of characteristics that evolution exploited in one platform and can not exploit in the second. The difference between the response of a SW evaluation and HW evaluation of two circuits described by the same chromosome may be caused by one or more factors originated either in the phenotype itself or in the way this is observed/evaluated. In some experiments, in particular when floating gate based solutions were involved (unusual for human designs), circuits may appear good during evaluation in the rapid sequence of tests of individuals in the population. However when the individual is evaluated alone, statically, it may not perform as well. (The following discussion refers to Fig 3, illustrating the evaluation paths in EEHW and IEHW). Some of these factors are summarized in the following:

1. Differences between model and real HW: a) Simplified models (e.g. to gain speed in SPICE runs), b) Incomplete models because of lack of information about fabrication, c) HW can change from the moment it was modeled/identified (temperature, radiation, operating conditions), d) HW can change in time after evaluation (e.g. slow discharge)

2. Simulator limitations (SW evaluation): a) Convergence conditions, which humans may be able to help by setting/adjusting values, b) Conditions

unknown *a-priori* (e.g. charges, initial conditions), in which case the system of differential equations can not be solved

3. HW testing limitations: a)Transients, b) Charge, e.g. remaining from a previously evaluated individual, c) Impedance loading of measured circuit, d) Time delays between physical signals (e.g. excitation) and outputs, e) Artifacts originating in signal generators, data acquisition paths, sampling, A/D, etc.

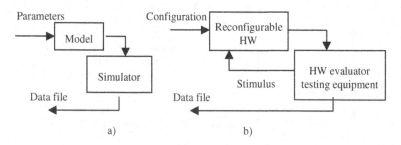

Fig. 3.   Path from chromosome to behavior data file a)extrinsic and b)intrinsic

The problems of portability to HW raises questions about the usefulness of the extrinsic solution from a practical/pragmatic/commercial point of view. On the other hand, not being able to simulate the solution evolved in HW largely diminishes the confidence in the intrinsic solution as it can not be analyzed and can not be proven or is not guaranteed to work outside the operating region used in the evaluations during evolution. When a solution exploits very specific effects, there may be situations where its applicability range may be very limited. This type of solution is a "point design", while what is needed is a "domain-wide" design, able to characterize a solution within a large envelope along several parameters: temperature, power supply, radiation effects, etc. In an attempt to achieve this type of robustness, it was proposed to evaluate each solution over a complete domain [4]. In our opinion the practicality of this approach is limited to simple cases due to cost issues, especially for space/military qualification. To deal with portability between two HW platforms, Thompson proposed to use different testing HW for different individuals in the evolutionary run. This can be done e.g. changing the area on the chip where the circuit is evaluated or moving between different chips, to ensure that only solutions that perform well on all platforms are selected during evolution. There is no guaranty that a newly evolved solution on one chip will work on others. One can not analyze/validate it in SW either if no accurate SW model is available (including parasitic effects that may be exploited by evolution as in [4]).

4   Mixtrinsic Evolution

Mixtrinsic evolution (ME) relates to evolving on mixed/heterogeneous populations, composed partly of models and partly of real HW [6]. This would constrain evolution to a solution that jointly simulates well in SW, and performs well in HW, i.e. a solution that exploits only the HW characteristics included in the SW model for producing the desired behavior (see Fig. 4). Solutions based on HW properties outside of the SW model are eliminated by evolution. In ME the population of candidate solutions is robust, more likely to be in agreement with common design rules, and, if novel, more likely to be patentable (i.e. to have generality and not depend on a fabrication process). The greatest advantage of the resulting solution is that it can both operate in HW and be analyzed in SW to explore its behavior outside the domain within which it was evolved. This is the only way to have insights and confidence in the evolved HW solution. Also, the resulting circuit is more likely to be portable to other HW platforms.

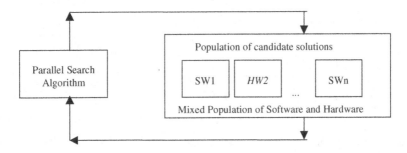

Fig. 4. Mixtrinsic EHW: evaluations of mixed populations comprised of both hardware and software solutions

Two types of ME are further detailed: *complementary and combined ME.* In *complementary* MEHW, candidate solutions are evaluated after being alternatively reassigned to either a HW or a SW platform (subject to random or deterministic choice). For example, an individual in a generation would have probability P to be evaluated in HW and probability (1-P) to be evaluated in SW. Assuming HW evaluates faster than SW one can speed-up evaluations by having a high value of P, which will cause a larger population to be evaluated in HW. The probability P, and related to it the ratio of individuals evaluated in HW over the total population, could also be variable parameters, adjustable during evolution.

In what we refer here as *combined* MEHW, each individual is evaluated both in HW and SW, and a combined fitness function is calculated. In the simplest case this can be a simple average of the two components or may involve adjustable weights etc.

We refer to the above description as a *matching* ME, for which the emphasis was on reinforcing the matching of similar characteristics of the SW models and the HW it describes. An opposite idea would be to reinforce dissimilarities and reinforce HW (or SW) distinctive characteristics, i.e. mismatches. We will refer to this as *mismatching* MEHW. This paper will demonstrate MEHW using a Field Programmable Transistor Array (FPTA) as evolutionary testbed.

5   FPTA Architecture

The FPTA was proposed as a flexible, versatile platform for EHW experiments and developed as an intermediate step toward a stand alone evolvable System-On-a-Chip (SOC) [7]. The architecture is cellular, and has similarities with other cellular architectures as encountered in FPGA (e.g. Xilinx 6200 family) or cellular neural networks chips. The main distinguishing characteristic is related to the particular definition of the elementary cell. This paper uses the first version of the FPTA cell, as illustrated in Fig. 5 (a new version is currently under development.) The structure is largely a "sea of transistors" where transistors are interconnected by other transistors that act as signal passing devices (gray-level switches). Details of the FPTA, its HW implementation and evolutionary experiments on FPTA can be found in [7]. The flexibility and versatility of the FPTA in implementing a variety of building blocks used in electronic circuits, as well as a discussion in the context of other programmable devices can be found in [8].

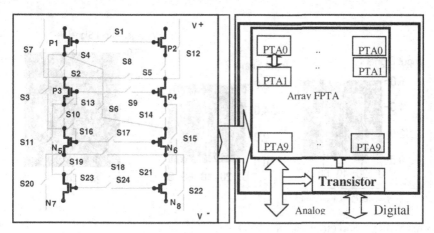

Fig.5 FPTA cell and FPTA chip

What is essential is that reconfiguration at transistor level allows definition of building blocks or subcircuits at a variety of levels of granularity. At lowest level one can configure subcircuits such as current mirrors and differential pairs, while more complex blocks such as logical gates, Operational Amplifiers (OpAmps), can also be easily configured. The level of granularity can be set by the designer, who can freeze the architecture to define high level components for evolution. Alternatively one can expect evolution to come with the building blocks that are the most suitable for the particular application (in the same sense as Koza's Automatic Defined Functions [4]).

The FPTA was exercised on a testbed that supports HW and SW evaluations (intrinsic/extrinsic). The SW subsystem makes use of the Caltech 256-processor HP Exemplar parallel computer to run multiple copies of SPICE. The HW subsystem is built around National Instruments LabView, associated data acquisition boards, signal generators, and other equipment, see [7] for more details.

6   Mixtrinsic evolution experiments: on convergent and divergent ME

This section exemplifies the portability problem and demonstrates the ME's ability to solve this problem. The following experiments are shown: a) Extrinsic evolution, with the resulting solution valid in SW but invalid when tested in HW, b) Intrinsic evolution, with the resulting solution valid in HW but invalid when tested in SW, c) Mixtrinsic evolution, with the resulting solution valid both in SW and HW.

The experiments show the evolutionary synthesis of an AND gate, using one FPTA cell. The input signals and an acceptable output response are shown in Fig. 6. The level 'high' input signals corresponding to logical '1' were controlled to keep their level for 5 ms, which corresponds to 20 samples on LabView graphs of acquired signal from HW. All experiments (about 20 runs for each case in a) and b) and 5 each for c1) and c2) below) used 30 individuals for 30 generations.

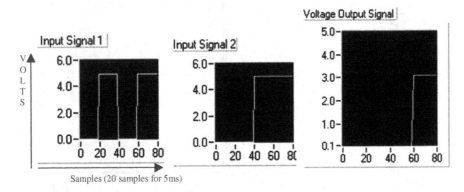

Fig. 6. Inputs and output for an AND gate

a) Extrinsic evolution. Two best individuals in the last generation are the solutions presented here for discussion. One of the solutions could be validated in HW. However, the circuit shown in Fig. 7, which is in fact the solution with the highest fitness, does not give satisfactory response when downloaded into HW. Thus, two direct observations can be made: a) one solution is validated in HW while the other is not, b) in this particular case, the "better " (in the sense of the fitness function that rewarded for higher value of the '1' level) solution in SW performs worse in HW. It appears that the solutions obtained through extrinsic evolution may not work in HW. Moreover, in many cases, there is no way to know for sure if it works without validating in actual HW.* (* We believe this reflects the current state-of-the-art, but admittedly we are strongly biased by our own experience with a certain model and HW. We believe that increasingly higher confidence in a solution would come from minimizing the negative effects of the factors discussed in Section 3. We also refer mainly to effects in analog circuits, and especially to those NOT relying on well understood building blocks, such as Op. Amps etc).

b) Intrinsic evolution. A circuit obtained intrinsically (best individual after a run with 30 individuals and 30 generations) and its response in HW and SW are shown in Fig. 8. The conclusion is that the solutions obtained through intrinsic evolution may not work in SW (see circuit responses in SW and in HW in the figure).

c1) <u>Combined Mixtrinsic Evolution (Matching)</u>. Each individual was evaluated both in HW and SW. The combined fitness was a simple average. The SW and HW responses are similar. The resulting solution is shown in Fig. 9.

c2) <u>Complementary Mixtrinsic Evolution (Matching)</u>. Each individual was allocated either to HW or SW evaluation with a 50% probability. The response of the resulting solution is identical to that illustrated in Fig. 9 (although the circuit is slightly different) and is omitted for space reasons.

In all experiments, the best 6 individuals of the last generation were tested both in HW and SW and they displayed similar responses. Although this is only empirical evidence, there is a good reason to believe that selection pressure would indeed favor solutions that display similar response in HW and SW.

d) <u>Divergent ME</u>. exploiting the distinctive characteristics of HW (or SW): Once accounted for the likelihood of obtaining mismatched responses between HW and SW, it appears straightforward to accept that selection pressure can force things in this direction (of mismatches). We are currently performing experiments in which we use combined evolution (each individual is evaluated twice, once in HW and once in SW). The combined fitness functions are either ratios of fitness of HW over fitness of SW, or its derivations as the sum of HW fitness and the inverse of SW fitness. Preliminary experiments illustrate that, indeed, resulting circuits produce the expected result in HW, while not being able to give a good response in SW.

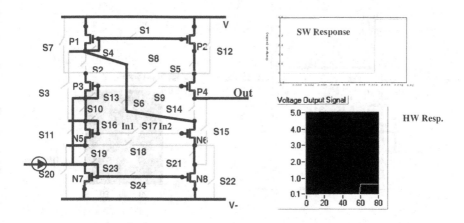

Fig. 7. Extrinsically evolved circuit, its response in SW and invalid response in HW

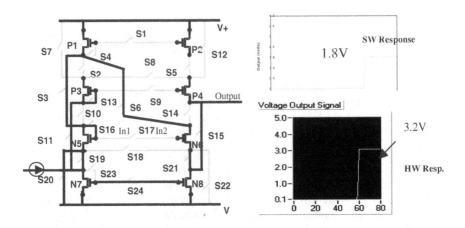

Fig. 8. Intrinsically evolved circuit, its response in HW and its invalid response in SW

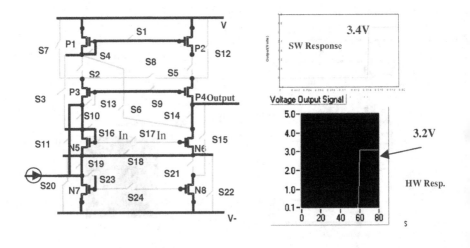

Fig. 9.   Circuit obtained by mixtrinsic evolution, its valid responses in SW and in HW

In the graphs of Figures 7, 8 and 9, the axis represent samples (X) and response in volts (Y).

7 Conclusion

Both Intrinsic and Extrinsic EHW appears to suffer from a portability problem (solutions evolved in SW do not run in HW and vice-versa), which is here illustrated through evolutionary experiments for the synthesis of an AND gate. A new approach introduced here and referred to as *mixtrinsic* evolution uses heterogeneous populations of individuals, some of which are evaluated extrinsically and some intrinsically. *Convergent mixtrinsic* evolution reinforces similarities between SW and HW behavior. Two flavors of the convergent style are demonstrated: complementary (population is mixed within the same generation, each individual being randomly evaluated either in HW or SW) and *combined* (each individual is evaluated both in HW and SW and a combined fitness is assigned). The demonstration uses Field Programmable Transistor Array architecture and shows that all the best individuals evolved in this way are validated both in HW and in SW. The opposite flavor of *mixtrinsic* evolution introduced here is *divergent* evolution, in which case selection rewards the distinctions between HW and SW, e.g. forcing circuits that exploit HW characteristics not modeled in SW.

Acknowledgements

The research described in this paper was performed at the Center for Integrated Space Microsystems, Jet Propulsion Laboratory, California Institute of Technology and was sponsored by the Defense Advanced Research Projects Agency (DARPA) under the Adaptive Computing Systems Program managed by Dr. Jose Munoz.

References

1. Thompson, A. Silicon Evolution. In: Proceedings of Genetic Programming 1996 (GP96), J.R. Koza et al. (Eds), pages 444-452, MIT Press 1996
2. Stoica, A., Keymeulen, D., Tawel, R., Salazar-Lazaro, C. and Li, W.   (1999) "Evolutionary experiments with a fine-grained reconfigurable architecture for analog and digital CMOS circuits. *Proceedings of the First NASA/DOD Workshop on Evolvable Hardware*, Pasadena, CA, July 19-21, IEEE Computer Society Press, pp. 76-84
3. Koza, J., F.H. Bennett, D. Andre, and M.A Keane, "Genetic Programming: Darwinian invention and problem solving", Morgan Kaufmann Publishers, San Francisco, CA, 1999
4. Thompson, A., Layzell, P. and Zebulum, R., Explorations in design space: unconventional electronics design through artificial evolution, IEEE Transactions on Evolutionary Computation, September 1999, V.3, N.3 pp. 167-196
5. Higuchi, T. et al., Real-world applications of analog and digital evolvable hardware, IEEE Transactions on Evolutionary Computation, September 1999, V.3, N.3 pp. 220-235
6. Stoica, A. Evolutionary design technique using a mixed population of candidate solutions, some of which are evaluated in SW and some in reconfigurable HW, New Technology Report NPO 20733, JPL, California Institute of Technology, July 1999
7. Stoica, A. Towards Evolvable Hardware Chips: Experiments with a Programmable Transistor Array. Proc.7th International Conference on Microelectronics for Neural, Fuzzy and Bio-inspired SYstems, *Microneuro'99*,. Granada, Spain, April 7-9, 1999, pp. 156-162
8. Zebulum, R. Stoica, A. and Keymeulen, D., A flexible model of CMOS Field Programmable Transistor Array targeted to Evolvable Hardware, to appear in the 3rd Int. Conf. on Evolvable Systems, ICES2000, Edinburgh, Scotland, April, 2000.

Evolution of Robustness in an Electronics Design

Adrian Thompson and Paul Layzell

Centre for Computational Neuroscience & Robotics, and
Centre for the Study of Evolution; School of Cognitive & Computing Sciences,
University of Sussex, Brighton BN1 9QH, UK
adrianth, paulla @cogs.susx.ac.uk,
WWW home page: http://www.cogs.susx.ac.uk/users/adrianth/

Abstract. Evolutionary algorithms can design electronic circuits that conventional design methods cannot, because they can craft an emergent behaviour without the need for a detailed model of how the behaviours of the components affect the overall behaviour. However, the absence of such a model makes the achievement of *robustness* to variations in temperature, fabrication, etc., challenging. An experiment is presented showing that a robust design can be evolved without having to resort to conventional restrictive design constraints, by testing in different conditions during evolution. Surprisingly, the result tentatively suggests that even within the domain of robust digital design, evolution can explore beyond the scope of conventional methods.

1 Rationale

Evolutionary algorithms can design electronic circuits that conventional electronics design methods cannot. Take a basic notion of an evolutionary process as the repeated action of selection upon individuals replicating with heritable variations [6]. Then an evolutionary algorithm can proceed by making undirected stochastic variations; a selection mechanism makes those variations that give rise to better observed behaviour more likely to persist, to be further embellished by subsequent rounds of variation and selection.

In contrast, every step in the derivation of a circuit design using conventional methods is taken with respect to some model of how it will affect final circuit behaviour. Even in a stage of iterative modification and testing, the alterations are chosen with reference to some model of their expected effect.

A circuit's behaviour may be termed *emergent* [3, 1, 2] if it cannot feasibly be predicted in detail given only a knowledge of the individual components and how they are interconnected. An evolutionary algorithm, needing no model to predict the effects of the component-level variations applied, can craft an emergent circuit behaviour, whereas conventional methods can not. The latter must constrain the components' interactions, designing within the subset of possible circuit structures and dynamics for which the necessary models are tractable.

Previous experiments have confirmed that an evolutionary algorithm can derive a circuit beyond the scope of conventional methods in this way [9, 13], but

J. Miller et al. (Eds.): ICES 2000, LNCS 1801, pp. 218–228, 2000.

raised the issue of *robustness*. Here, robustness is the ability of a circuit to operate adequately over a specified range of variations in temperature, fabrication, power-supply voltage, and so on. Notice that these perturbations are often at the level of physical components. The constraints of conventional design, in allowing models of how the components affect the overall behaviour, also simplify the achievement of robustness. How much robustness is required is part of the specification of an application, but some is nearly always necessary.

The main benefit of allowing evolution to explore designs without constraint is that the circuits produced can be different to those otherwise obtainable, e.g. [5]. This novelty means that the circuits might be better in some circumstances, for example being smaller, more power efficient, or displaying graceful degradation. This paper reports an experiment showing that evolutionary design can be induced to produce a robust circuit, without having to impose the customary constraints on what designs can be explored. The paper finishes with a discussion of the wider implications of the result.

2 Experiment

The circuit to be evolved has one input and one output. As the input is stimulated with a square wave of either 1kHz or 10kHz, the output is to give a steady high voltage for one input frequency, and a steady low for the other. This frequency discrimination task was used in the earlier experiments, and was chosen to be as simple as possible while raising the issues of interest. In particular, some sort of stable dynamics is needed within the circuit. The tone discrimination task can also lead the way to more sophisticated cases of pattern recognition over time, such as word recognition [11]. We seek to explore new means to generate robust behaviour, which does not necessarily imply that the behaviour could not be achieved in some other way using conventional methods.

The circuit was evolved to be a configuration of the now-obsolete Xilinx XC6216 field-programmable gate array (FPGA). Each candidate design (evolutionary variant) was tested by configuring four XC6216 chips having the conditions shown in Table 1. Evaluations at the target task were then performed on all four chips simultaneously and independently. The evolutionary fitness of the candidate was taken to be the worst of the four measurements. This multi-FPGA, multi-condition apparatus, called 'The Evolvatron', was fully described in [12].

Only the configuration of a 10×10 region of the 64×64 array of cells on the FPGA was subject to evolution. For each chip of Table 1, the 10×10 region was translated in position by a different but constant amount, as shown. Only nearest-neighbour interconnections between cells were enabled. With reference to the datasheet [15], this leaves 10 multiplexers ('muxes') to be configured in each of the 100 cells: four 4:1 neighbour muxes, three 4:1 'X' muxes to select the inputs to the cell's function unit; and within the function unit, two 4:1 muxes (Y1 and Y2), and the single 2:1 CS mux selecting either a combinatorial or a

Chip	Fabrication	Package	Interface	Temperature	PSU	Output load	Position
1	Seiko	PQFP	parallel	in PC	PC's 5V s.m.	-	(37,30)
2	Yamaha	PLCC	serial	ambient	5V lin.	1kΩ	(32,0)
3	Yamaha	PGA	serial	60°C	4.75V s.m.	-	(63,0)
4	Seiko	PGA	serial	−27°C	5.25V s.m.	-	(37,54)

Table 1. Conditions of the chips used for fitness evaluation. 'Temperature' is that of the surface of the chip package. See [14] for the thermal properties of each package. Power supply units (PSUs) were either switched-mode (s.m.) or linear regulated (lin.). 'Position' gives the (row, column) location of the north-west corner of the 10×10 circuit within the 64×64 array, with (63,0) being the extreme north-west corner of the array.

sequential output. The RP (register protect) mux, which can be set to prevent the cell function unit's flip-flop from ever changing, was always set to OFF.

In contrast to the earlier experiments, a clock signal was supplied to all of the FPGAs, and was the clock input to the rising-edge triggered D-type flip-flop inside every cell's function unit. The clock source was a single external crystal oscillator, nominally 6MHz. Under evolutionary control, via the Y1, Y2 and CS muxes, each cell's function unit could be synchronous to the clock (function output taken from the flip-flop), asynchronous to the clock (flip-flop output not selected), or mixed (function output is a combinatorial function of the possibly-asynchronous inputs as well as of the flip-flop's output). Hence, there is no constraint that evolution must produce a synchronous digital design: the clock is a stable dynamical resource to be used or ignored as selection deems fit, rather than being an enforced synchronisation signal [10]. No design rules were imposed.

The evolutionary algorithm was a (1+1) Evolution Strategy (ES) [7,8]. A mutation operator was defined to select one of the 100 cells at random, then one of that cell's 10 muxes at random, and then to randomly reconfigure that mux to select a different input. In the ES, each new variant was produced by applying the mutation operator three times to the parent.

With a new variant configured to the FPGAs, the inputs were stimulated with a sequence of 100ms bursts of 1kHz and 10kHz tones. The tones were shuffled into a random order for each trial, such that there were an equal number of each frequency. The single source of the tones was a 1.000000MHz crystal oscillator module, followed by hardware dividing the frequency by either 100 or 1000.

Separate analogue circuits integrated the output of each chip over each 100ms tone, giving a value proportional to the average output voltage during that time. Let the integrator reading of FPGA chip c at the end of test tone number t be i_t^c ($t=1,2,\ldots T$). Let S_1 be the set of 1kHz test tones, and S_{10} the set of 10kHz test tones. Then the evaluation of the performance of the configuration on chip c was calculated as:

$$E^c = \frac{1}{2T} \left| \sum_{t \in S_1} i_t^c - \sum_{t \in S_{10}} i_t^c \right| - P^c \tag{1}$$

and the fitness F was $\text{MIN}_{c=1}^{4}(E^c)$. The term P^c was used towards the end of the run to penalise solutions having spikes or glitches in their output. For each FPGA, a separate resetable counter made from 74LS devices counted the number of low$\Rightarrow$high logic transitions on that FPGA's output during each test tone. If the number of low$\Rightarrow$high transitions during tone t was $N^c(t)$, then a counter would return the value $R^c(t) = \text{MIN}(\ N^c(t),\ 255\ )$. Then P^c was defined as

$$P^c = w \times \frac{1}{2T} \sum_{t=1}^{T} \text{MAX}(\ R^c(t) - 1,\ 0\ ) \tag{2}$$

where w is a weighting factor, initially zero.

The ES algorithm proceeded as follows, where L is an integer initially 1 but later increased to lengthen the trials:

- Download the initial parent configuration to the FPGAs.
- Measure F_{parent} over $50 \times L$ tones.

REPEAT

    - Generate three mutations (the variation operator) and
      partially reconfigure each FPGA with all three.

    - † Measure $F_{offspring}$ over a total of $24 \times L$ test tones,
      aborting after $8 \times L$ or $16 \times L$ tones if $F_{offspring} < F_{parent}$.
      Before the final group of $8 \times L$ test tones (if reached), all FPGAs are
      hardware-reset and completely reconfigured from scratch with the
      new variant.

    - IF ($F_{offspring} \geq F_{parent}$)
      The offspring becomes the new parent.
      All three mutations are left in place.
    ELSE
      The parent is left unchanged:
      partially reconfigure the FPGAs to undo the three mutations.
    END IF

    - IF (15 offspring have been generated without any replacing the parent)
      Hardware reset the FPGAs & reconfig. from scratch with the parent.
      Extend the parent's evaluation by a further $10 \times L$ test tones.
      Wait a further 15 offspring before doing this again.
    END IF

END REPEAT

The initial parent was found through random search for a starting point giving better than trivial performance. Partial reconfiguration is used above because some of the FPGAs are reconfigured via a relatively slow serial interface. The simple variable-length evaluation scheme copes with measurement noise and the stochastic behaviours of some candidate configurations. Termination was by a human observer. The algorithm relies on the existence of pathways of change that make no immediate difference to the fitness, but that alter the opportunities for later improvement. See [4] for a discussion of such 'neutral networks' in the context of evolutionary electronics design. One reason for choosing such a simple algorithm was to simplify the study of the role of neutrality, to be reported elsewhere.

The initial parent configuration ($F=0.43$ with $w=0$) was found after testing 75679 randomly generated individuals, using only the fitness measuring step marked † above (with $L=1$), and with a hardware reset and complete reconfiguration of the FPGAs between individuals. In the main phase of evolution ($L=1$, $w=0$), 25000 triple-mutations were accepted out of 861348 attempted (although most evaluations would be aborted early) resulting in a circuit with fitness $F=6.17$. Then w was increased to $1.4/255$ and a further 10 triple-mutations were accepted out of 265 trials. Then with $L=2$, $w=1.0$, a further 870 out of 32338 triple-mutations were accepted. Finally, with $L=5$, $w=5.0$, a further 3000 out of 131005 mutants were accepted. Each of these latter phases resulted in a reduction in the number of unwanted spikes in the output, eventually eliminating them entirely.

3   Results and Preliminary Analysis

The final circuit works near-perfectly on all of the FPGAs and conditions of Table 1. The only room for improvement is that sometimes when the input frequency is changed, the circuit's output responds after several cycles of the input, rather than after the minimum of a single half-cycle. For steady 1kHz or 10kHz input, in 24-hour tests no unwanted transitions in the output occurred.

As a further test of robustness, the final circuit was tested on six XC6216 FPGAs that were not used during evolution, including samples from both foundries. Each chip was first frozen to $\simeq -50°$C using freezing spray, then the circuit's operation was monitored as the chip warmed to room temperature. The same near-perfect performance as before was seen in all of these conditions, indicating that the evolved design is usefully robust.

Figure 1 shows the functional part of the circuit. The configuration can be changed to clamp to constant values those cells and connections not shown, without affecting the behaviour.

We now wish to discern whether there has been any advantage in allowing evolution the freedom to explore beyond conventional digital design rules. To see if any or all of the circuit's behaviour fits into a digital-logic model of operation, it was simulated in a well-known digital simulator (PSpice^{TM}). Working with noise-free binary signals, for each kind of component there is a model describing the internal propagation times from the component's inputs to its output. At the start of the simulation, the propagation times of each component are increased by a load model, to take account of what loads that output must drive.

The simulation model of the FPGA used components from the 74AC family, with the propagation times changed to match the approximate figures given in the datasheet [15]. Where necessary, the simulator defaults for the relationships between minimum, typical, and maximum timings were used. The other details, such as the loading model, were unchanged. The reconfigurable multiplexers

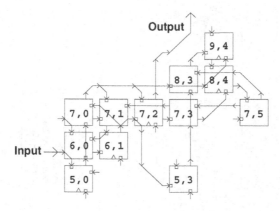

Fig. 1. The functional part of the evolved configuration. The large boxes represent the cells. A wire shown originating from the perimeter of a cell is the output of that cell's function unit. Inputs to a cell's function unit are indicated with small squares. If a cell's function unit output is taken from it's flip-flop (synchronous), then a small triangle is drawn at the bottom of the cell. The configurations of the function units are not shown. The cells are labelled with their (row, column) co-ordinates within the 10×10 evolved array.

were replaced by inverters or buffers hard-wired according to the final evolved configuration. The resulting model is shown in Figures 2 and 3, which also show all the details not visible in Figure 1.

The circuit worked first time in simulation, without any need to fine-tune the component timings. This means it does not rely on any aspects of the chips not captured in the simulation. In particular, it does not rely on any analogue effects, in contrast to an earlier experiment in which robustness was not an objective [13]. On a Pentium 233MHz processor with 32MB of memory, the simulation ran $\simeq 62000$ times slower than real time.

Using the schematic and simulation traces, the circuit appears to operate as shown in Figure 4. First, the input I is 'retimed' (sampled) to the 6MHz clock in cell (5,0). When the retimed input R is high, cell (7,0) toggles at the clock frequency. When R is low, this oscillation stops. Since the sampling in cell (5,0) delays R by up to (but less than) a clock period with respect to I, the number of times the oscillator toggles is completely determined by how long the raw input is high, and hence on the input frequency. For some input frequencies, the oscillator toggles an odd number of times, so finishes in a different state, whereas for others it does not. This is the heart of the timing mechanism; cell (7,2) simply holds the final value of the previous oscillation while the next one is going on, and this is the output of the circuit.

The core mechanism described so far uses only three cells, but would, at best, produce a constant output for one input frequency and an output toggling every cycle at the other input frequency. The function of all of the other cells in the circuit is to correct this, by delaying the retimed input. The delay on

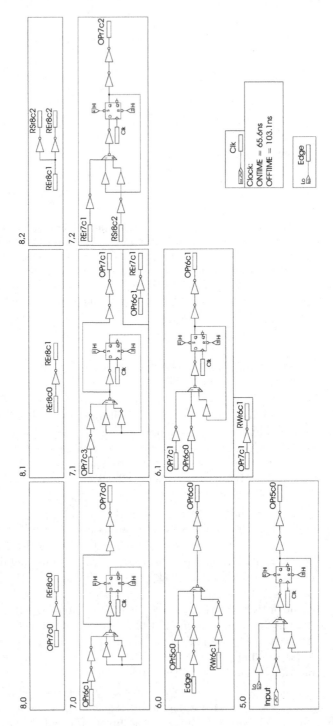

Fig. 2. The left half of the simulation model.

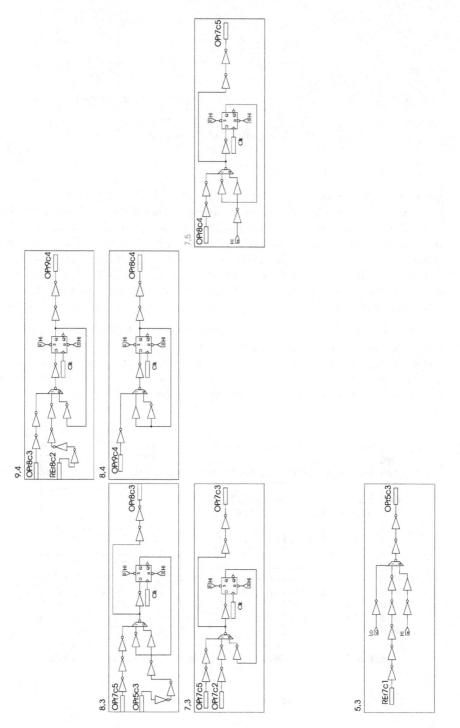

Fig. 3. The right half of the simulation model.

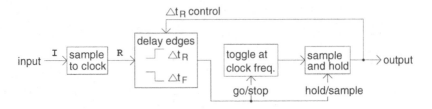

Fig. 4. An outline of the mechanism.

falling edges is constant, but for rising edges is variable, and is a function of the present output of the circuit. This is arranged so that if the input is 1kHz while the output is high, or if the input is 10kHz while the output is low, then the oscillator will have time for an odd number of toggles in a high half-cycle of the input, and the output will change state. If the input is 1kHz while the output is low, or 10kHz while it is high, then the oscillator toggles an even number of times and the output is constant. The implementation of the variable delay is not yet understood.

The simulator is able to report whenever there is a flip-flop data-to-clock setup time violation: that is, when the input to a flip-flop changes so soon before the clock edge arrives that the resulting output is uncertain. Setup time violations are only ever reported in the cell (5,0) which retimes the input: the other nine flip-flops in the evolved design operate in a reliable clocked fashion.

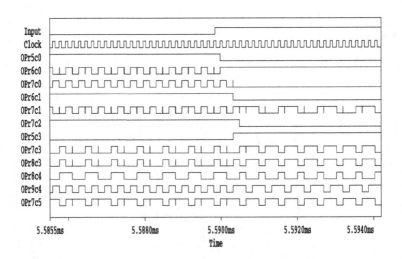

Fig. 5. Typical cell outputs during simulation.

4   Discussion

The observed robustness, the success of the simulation, and the absence of setup time violations, together provide good evidence that the circuit operates as a well-behaved digital design. However, the simulation waveforms (Figure 5) show many spikes and glitches of various durations that are determined by analogue time delays through the circuit. Such dynamics are avoided in digital design as leading to unreliable or non-robust operation. Presumably the present design evolved to be insensitive to these, perhaps completely ignoring them.

Together, these findings give a fascinating result. The circuit has evolved to achieve robustness by finding a well behaved, clocked, digital mode of operation. However, it is not at all clear what digital design rules could have been formulated in advance, that would have guaranteed robust digital operation and yet permitted this evolved circuit.

A surprising hypothesis is suggested: Even within robust digital design, unconstrained evolution can produce circuits beyond the scope of conventional design rules. Previously it had been assumed that the domain of robust digital design was fully covered by conventional design rules, and that novel territory for evolutionary exploration must lie in other domains [13]. Given the undoubted utility of robust digital designs, it is an exciting possibility that evolution could explore novel regions of design space, containing circuits that may be *better* for some applications.

More work is needed to investigate this bold idea, and the results presented here do not yet conclusively prove it. Future work should also look at the characteristics of the evolutionary intermediates that aided the discovery of the final solution. Finally, it should be noted that in nature there are numerous strategies for robustness other than clocked digital design [10], and these may be of interest to evolutionary electronics when using analogue components.

Acknowledgement

to Phil Husbands, Inman Harvey and Ricardo Salem Zebulum. This work was primarily funded by EPSRC, and we also thank the following for support: Xilinx, British Telecommunications, Hewlett-Packard, Zetex, and Motorola.

References

1. P. Cariani. Emergence and artificial life. In C. G. Langton, C. Taylor, J. D. Farmer, and S. Rasmussen, Eds, *Artificial Life II*, pp. 775–797. Addison Wesley Longman, 1992.
2. J. P. Crutchfield and M. Mitchell. The evolution of emergent computation. *Proc. Nat. Acad. Sci. USA*, 92(23):10742–10746, 1995.
3. S. Forrest. Emergent computation — self-organizing, collective, and cooperative phenomena in natural and artificial computing networks. *Physica D*, 42(1-3):1–11, 1990.

4. I. Harvey and A. Thompson. Through the labyrinth evolution finds a way: A silicon ridge. In T. Higuchi and M. Iwata, Eds, *Proc. 1st Int. Conf. on Evolvable Systems (ICES'96)*, vol. 1259 of *LNCS*, pp. 406–422. Springer-Verlag, 1997.

5. J. R. Koza, F. H. Bennett III, M. A. Keane, et al. Searching for the impossible using genetic programming. In W. Banzhaf, J. Daida, A. E. Eiben, et al., Eds, *Proc. Genetic and Evolutionary Computation conference (GECCO-99)*, pp. 1083–1091. Morgan Kaufmann, 1999.

6. J. Maynard Smith. *Evolutionary Genetics*. Oxford University Press, 1989.

7. I. Rechenberg. Cybernetic solution path of an experimental problem. Royal Aircraft Establishment, Library Translation 1122, 1965. Reprinted in 'Evolutionary Computation — The fossil record', D. B. Fogel, Ed., chap. 8, pp. 297-309, IEEE Press 1998.

8. H.-P. Schwefel and G. Rudolph. Contemporary evolution strategies. In F. Morán, A. Moreno, J. J. Merelo, and P. Chacon, Eds, *Advances in Artificial Life: Proc. 3rd Eur. Conf. on Artificial Life*, vol. 929 of *LNAI*, pp. 893–907. Springer-Verlag, 1995.

9. A. Thompson. Silicon evolution. In J. R. Koza, D. E. Goldberg, D. B. Fogel, and R. L. Riolo, Eds, *Genetic Programming 1996: Proc. 1st Annual Conf. (GP96)*, pp. 444–452. Cambridge, MA: MIT Press, 1996.

10. A. Thompson. Temperature in natural and artificial systems. In P. Husbands and I. Harvey, Eds, *Proc. 4th Eur. Conf. on Artificial Life (ECAL'97)*, pp. 388–397. MIT Press, 1997.

11. A. Thompson. Exploring beyond the scope of human design: Automatic generation of FPGA configurations through artificial evolution (Keynote). In *Proc. 8th Annual Advanced PLD & FPGA Conference*, pp. 5–8. Miller Freeman, 12th May 1998. Ascot, UK.

12. A. Thompson. On the automatic design of robust electronics through artificial evolution. In M. Sipper, D. Mange, and A. Pérez-Uribe, Eds, *Proc. 2nd Int. Conf. on Evolvable Systems (ICES'98)*, vol. 1478 of *LNCS*, pp. 13–24. Springer-Verlag, 1998.

13. A. Thompson, P. Layzell, and R. S. Zebulum. Explorations in design space: Unconventional electronics design through artificial evolution. *IEEE Trans. Evol. Comp.*, 3(3):167–196, 1999.

14. Xilinx. Packages and thermal characteristics V1.2, August 1996. In *The Programmable Logic Data Book*, chapter 10. Xilinx, Inc., 1996.

15. Xilinx. XC6200 field programmable gate arrays. Data Sheet, Xilinx, Inc., April 1997. Version 1.10.

Circuit Evolution and Visualisation

Peter Thomson

School of Computing, Napier University,
Edinburgh, Scotland, UK.
Email: p.thomson@dcs.napier.ac.uk
Tel: +44-(0)131-455-4245

Abstract. The evolution of digital circuits up to the present time has concentrated from necessity upon small systems consisting of very few individual gates. In this paper, the potential for evolving larger systems more quickly is explored via a method of visualising the sub-components of the final network as and when they appear. It is envisaged that by doing this one will be able to take partially evolved solutions from short runs, then feed these to a further evolution. Therefore, instead of long random evolutionary runs, we end up with an approach consisting of a number of much shorter runs with evolution being assisted by sub-components that have already been identified. It is found that doing this does, indeed, speed the process in both the experiments described herein and in those of a companion paper [11].

1. Introduction

The current interest in the design of electronic circuits by artificial evolution is well documented, and particularly highlighted in the work of Koza [1], Thompson [2][3] and various others [4][5][6]. My own work in this field has concentrated mainly on the direct evolution of combinational logic circuits on either a free netlist framework [7], or on a simulation of the Xilix 6216 Field-programmable Gate Array (FPGA)[8]. This has proved relatively successful for small circuit designs, and particularly those which implement arithmetic functions such as adders and multipliers.

In this paper, the evolution of similar designs is investigated further by means of visualisation in an attempt to understand the underlying process of circuit/network development. In other words, by visualising the circuit netlists at various stages of development, it is hoped that we will be better able to understand which sub-blocks (circuit gates or functional cells) evolve first, and to then use that knowledge to accelerate the overall process.

The following sections give a brief description of the problem studied for the purpose of this analysis, the netlist framework being used to perform the evolution and also the nature of the subsequent visualisation. The results section then describes the various experiments carried out to ascertain early sub-system development, and the attempt to feed this information back into the evolutionary cycle to speed up the emergence of correct and finalised designs.

J. Miller et al. (Eds.): ICES 2000, LNCS 1801, pp. 229–240, 2000.
© Springer-Verlag Berlin Heidelberg 2000

2. The Two-Boxes Problem

The experiments, which are described here, for the purpose of analysing the way in which cellular networks evolve have been carried out on a test problem known as the *two-boxes* problem. This is a benchmark problem often used to test the effectiveness of algorithms in Genetic Programming (GP) and, as such, is described in Koza [10]. It consists of finding the relationship between a set of six input variables and an output variable. The six inputs are the dimensions – length, width and height – of two cuboid boxes, and the output, D, is the difference in their volumes. Hence, the formula that should emerge is:

$$D = L_0 W_0 H_0 - L_1 W_1 H_1 \tag{1}$$

Where L_0, W_0, H_0, L_1, W_1 and H_0 are the dimensions of the two boxes and D is the difference in their volumes.

The network is then effectively required to evolve this relationship (with a fitness which is the mean-squared error between the actual function outputs and those generated by each network) using only the integer operations +, -, * and /. This is directly analogous to evolving logic functions where a truth table of inputs related to desired outputs is presented, alongside a set of appropriate functional operations attributed to each individual cell. The reason that the two-boxes problem has been used here is because experiments are currently underway to compare traditional GP methods of solving this type of problem with an evolvable hardware approach. Comparative results, in terms of computational effort required, have proved favourable [12]. It is anticipated that the methods described in this paper will be fully applicable to logic design problems such as 2-bit or 3-bit multiplier circuits.

In section 4, experimental results are presented which deal initially with tests on this problem. Then a second problem is considered using the derived technique. This problem, which, although using similarly structured relationships (in terms of the numbers of inputs, outputs and internal operations), have been randomly generated and, unlike the two-boxes problem, were not known to the author before the solution network is evolved. It was felt that it was important to demonstrate the effectiveness of being able to visualise networks as they evolve, and to be able to influence the way in which that evolutionary process should develop. Using randomly generated problems, in blind tests, helps to show that the approach is generally applicable, at least to problems of a similar type to the one originally considered.

3. The Free Netlist Framework and Its Visualisation

Of the two approaches to circuit evolution that both myself, and my colleagues, have attempted [8], the free netlist approach has been the most successful in producing 100% functionally correct logic circuits. The Xilinx simulation, although based upon the internal structure of an actual device, has proved too restrictive in its routing capabilities to be able to offer the type of platform required for the artificial evolution of circuit designs. We have previously commented on this [8] by direct comparison of the two approaches, and argue that the architecture of physical devices for circuit evolution is still an area that needs thorough investigation. Therefore, it is the free netlist method that is used throughout the investigations described here.

The free netlist approach uses primarily two-input, single output cells, which combine the input values using some pre-defined functionality, to form the cell output value. The evolution of these structures is performed by a genetic algorithm (GA), which uses integer chromosomes, which completely define a circuit connectivity and functionality. Figure 1(a) below shows a typical chromosome, whilst Figure 1(b) shows the resultant circuit as produced by the visualisation software.

4 3 2 0 2 2 5 4 3 6 3 0 3 3 2 1 7 2 9 8 0 11 6 1 11 9 1 14

(a)

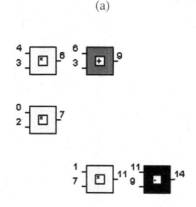

(b)

Fig. 1. A chromosome and resultant cellular network

Here, the GA evolves the circuits - albeit without the use of crossover, which has no discernible effect in this application - and additional software is then used to visualise the final cell structure and functionality. Redundant cells, i.e. those which are not physically connected - and, therefore, have no effect on the functional output - are optionally removed within this representation.

The notion of visualising the circuits, which emerge from the evolution, may then be used to investigate which parts of circuits evolve first. It is hoped that by doing this, one may be able to speed up what can prove to be a time-consuming procedure.

Firstly, a short run of the GA (over relatively few generations) is carried out, and a visualisation of the resulting circuits presented. The software examines individual generations for improvement in fitness over the run, and where this occurs, the chromosomes concerned are recorded to a file. A *fitness threshold* is used throughout the run of the GA to prevent the recording of improvements whilst the fitness is still very poor. This is done because there tend, in a random run of the GA (i.e. one with a randomly selected population), to be very many improvements early on, and experimental results have shown that the evolution of these cellular structures do not really provide consistent cell usage until the fitness falls to below a certain level.

Once the random run is complete, the software is then used in an alternative mode as a *results viewer*, and, at this stage, the user is able to examine those chromosomes that have represented an improvement in fitness over the run. Figure 2 shows a typical set of improvements for the two-boxes problem for 10 runs of the GA over 50 generations (the population size being 10, with a mutation rate of 5%).

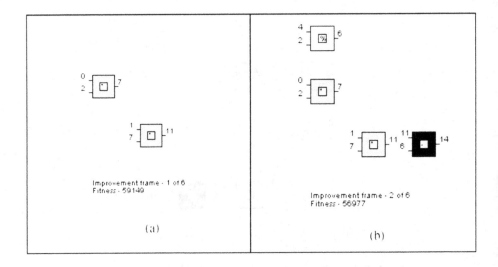

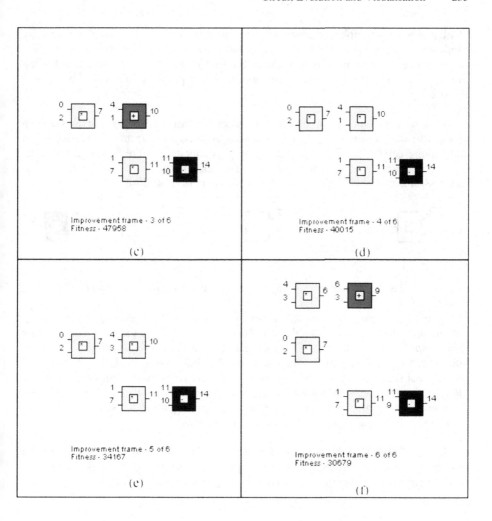

Fig. 2. A chromosome and resultant cellular network

Each improvement is examined by the user in order to determine which parts of the cellular structure are consistently present. The idea being that this will give insight into the final, solution network. In other words, looking at the example of Figure 2, it may be seen that certain cells are ever present within the structure, so one is led to suspect, that since the overall fitness is continually improving, that these cells will play some part in determining the 100% functionally correct solution. The idea is that these structures are genetically *related* to the solution structure, and so in a very short run of the GA we have been able to determine at least a part of the solution.

Further, since it is believed that these partial solution structures represent a genetic route to finding actual solutions, we should not waste this genetic material, but should, as it were, recycle this to assist in finding further improvements. This may be achieved by selecting particular cells within the improvement visualisations to be *locked down* i.e. completely preserved in terms of both functionality and connectivity and presented as part of a *pre-seeded population* to a subsequent run of the GA. Experiments suggest that this is an extremely fruitful procedure with regard to finding solutions to both the two-boxes problem and other problems for which blind tests were conducted.

With regard to the actual structure of the cell netlist used, it has previously been found [9] that a *single row, many columns* geometry is the most effective when attempting cell-based circuit evolution using this model. Figure 3 demonstrates the typical appearance of this structure type. The reason that this particular form is so successful is probably due to the *feedforward* nature of these designs.

Fig. 3. The single row, many columns geometry for cellular evolution.

In this arrangement it is possible for every cell input to be determined by the output of any preceding cell in the single row. This is not necessarily the case when *a many rows, many columns* geometry, such as that of Figure 2, is used. With this approach, the search space of possible solutions is much more attuned to finding the type of design that is ultimately desired. One which is highly feedforward, but with the ability to create appropriate sub-functions at cell outputs, which may then be used (and even re-used) at subsequent cell inputs. It is this particular quality that seems to give this structure an advantage over the two-dimensional forms.

In the following section, results of experiments are presented which show how the visualisation of cell structures, and the appropriate locking down of useful cells, is able to improve the overall evolutionary performance of this technique.

4. Results

Two main experiments were conducted. The first attempts to ascertain which cells within the final structure evolve first in the hope that this may provide an insight into the manner in which subsequent evolution will proceed in finding a functionally correct solution. The second then attempts to determine a procedure whereby a user of the system may quickly establish a partially evolved solution, lock down the relevant useful cells, and then proceed with the search in a more directed manner. In all of the experiments, the GA uses a population size of 10 and a mutation rate of 5% as this has previously been found to be successful in evolving logic circuits. Crossover is not used. Experiments using crossover have consistently revealed that, within this problem domain, the GA performs no better when crossover is used than when it is not used. Therefore, since its presence is slowing down program with no perceived benefit, it is removed from the runs of the algorithm. Strictly speaking, therefore, the algorithm used to evolve these structures is an *evolutionary search* based upon selection, mutation and replacement, rather than a GA, which normally implies the use of crossover.

In the first set of experiments conducted using the two-boxes problem, a two-dimensional geometry was used initially. However, it was later found that a more effective partial solution could be found using the single row, many columns model described previously and illustrated in Figure 3. Early experiments found that over a relatively short run of the algorithm, say 50 generations, cells would emerge which found a partial solution to the problem by completing one of the multiples e.g. $L_0 W_0 H_0$ or, more rarely, $L_1 W_1 H_1$. Therefore, structures such as that of Figure 4 were typical.

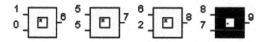

Fig. 4. Early stage evolution of partial solution to two-boxes problem.

This meant that within a fairly short run of the algorithm, useful genetic material was already emerging. However, not enough to be able to find a 100% correct solution. Typically, this required a much longer run of the algorithm, usually in the region of 1000 generations.

It was also found that if the fitness of chromosomes reached a particular threshold level (of the order 100,000), then this meant that the material that was emerging would be useful. If the fitness was not able to achieve this level, then there was no consistently emerging structure. This was borne out by examination of results files which showed improvement beyond the threshold as opposed to those that did not. Those that displayed such improvement always contained cells which would remain unaltered (or virtually so) over a number of improvement stages indicating that these cells were important to the overall fitness of the structure. Therefore, algorithm runs which passed the fitness threshold would typically display the type of improvement shown in Figure 4 over a number of results slides. One could then be confident that cells which remained unaltered over a number of improvement stages would be useful candidates to preserve within future runs of the algorithm. Figure 5 contrasts two runs, the first which attains the threshold level, and the second which does not, showing that the structure which reaches the threshold does indeed contain a partial solution ($L_0 W_0 H_0$), whilst the other has no readily identifiable useful genetic material.

When this preservation of identified useful cells was attempted initially, however, the results were disappointing. It was found that merely preserving the cells which formed the partial solution was, in itself, not sufficient to promote further improvement, but that one had also to preserve the *routing* information which connected these cells to the output. This meant also preserving the output cell to maintain the evolved cells' influence over the achieved fitness. When this was done, it was found that additional improvements were then possible. Further experimentation with this technique revealed that not only was the preservation of the output cell an important feature, but also its position within the structure. The three separate runs which produce the partial solutions of Figure 6 clearly show that this is the case. In each individual run, the algorithm is run 10 times over 50 generations to produce these results.

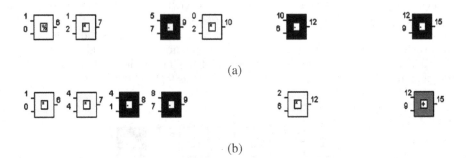

(a)

(b)

Fig. 5. Two partial solutions which contrast the difference in reaching the identified fitness threshold.

Table 1, then shows the achieved fitnesses for a further 10 runs of 200 generations for each of these partial solutions with cells from each preserved and used to seed the initial populations of the subsequent runs.

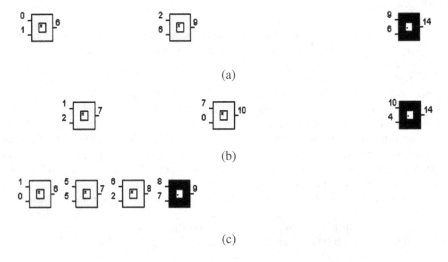

(a)

(b)

(c)

Fig. 6. Three independent partial solutions generated by runs of 50 generations.

Table 1. Results from seeding with the partial solutions from Figure 6.

Run	Fitness Improvement Seeded from Network 6(a)	Fitness Improvement Seeded from Network 6(b)	Fitness Improvement Seeded from Network 6(c)
1	34159	33569	33569
2	33569	33569	34159
3	34159	0	34159
4	0	34159	34159
5	45440	45440	34159
6	39662	0	34159
7	20637	29225	33569
8	0	220	34159
9	45440	48944	34159
10	33569	0	34159

The important point here is that whilst the networks of Figure 6(a) and 6(b) were able to produce 100% correct solutions (those with fitness 0) within what is effectively 250 generations – a factor of 4 improvement on the GA with a randomly selected population (which typically requires runs of 1000 generations). However, the run of 6(c) was not able to find a solution. The reason would appear to be that the output cell in 6(c) is physically located in too close a proximity to the cells which form the $L_0W_0H_0$ factor. This leaves no room for expansion of the other cells required to form the second factor and, thus, the solution unless radical mutation takes place to remove the original output cell, re-locate it, and re-connect it to the existing cells. The algorithm finds this much more difficult than using an already conveniently located output cell as occurs in the first two examples.

To test this notion further, the software was modified to enable preservation of the output cell, but to move it physically to a location more in keeping with the structure of 6(a) or 6(b). The modification is shown in Figure 7. When this was done, and the algorithm is run with exactly the same parameters, a correct solution was found on two occasions.

Fig. 7. Modified version of circuit 6(c).

In order to test further whether or not this technique can be extended to general problems - and is not merely successful when the solution is known a priori - a small program was created to generate random test problems whose solutions are hidden. This software was set up to generate the test table for the problem in one file, whilst generating the relevant equation in a separate file. Using the same approach of evolving a partial solution over a short run, then locking down cells which appear to contribute over a number of improvements, and then evolving over a slightly longer run, it was found that the solution could virtually always be determined by GA. Figure 8 shows one particular example. If we consider the six inputs to be designated *a, b, c, d, e* and *f*, then the equation which has evolved is:

$$ae^2 + (1+e^2) \tag{2}$$

and this was, in fact, the randomly chosen blind test for this problem. All other runs were equally able to produce a correct solution.

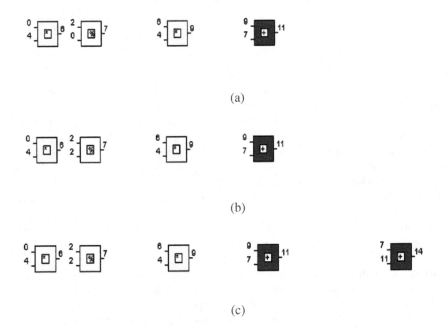

(a)

(b)

(c)

Fig. 8. 8 Evolution of a randomly generated equation using visualised cell improvement

5. Conclusions

This paper has described attempts to improve the speed of evolution of evolvable hardware systems by visualising which sub-components (or cells) are first to evolve within the process.

The technique was initially examined by consideration of the two-boxes problem which is a relatively simple GP benchmark. The reason for this is that it is simple to implement, and is directly analogous to evolving logical truth tables for digital circuit design. However, perhaps more importantly, there is a lot of available data relating to attempts to evolve this solution with which the present technique may be compared. This is done in [12]. The aim here was to create a method whereby cells which evolved early, to form reasonable partial solutions, could then be maintained into future evolutionary runs. It was found that by doing this 100% correct solutions were arrived at more quickly. However, certain conditions had to be met beforehand. These were: (a) that any connection between the influential cells of the partially evolved solution and the output had to remain intact, and (b) that the relative positions of cells within the overall geometry is important - cells which are too closely packed, for example, present too great a constraint to the subsequent evolution.

This meant that a general procedure emerged from the experiments that were constructed. This is:

Step 1: Perform a short run of the GA (~50 generations).

Step 2: Identify those cells which are influential, and therefore represent a partial solution.

Step 3 : Preserve these and execute the GA for a longer run (~200 generations).

Step 4 : If no solution found, return to step 2 and repeat.

Using this approach, in this small problem, a factor of 4 improvement was achieved on the number of generations over which the GA had to be run in order to find 100% functionally correct solutions.

As a further test, this method was applied to previously unseen problems whose size (in terms of the numbers of operations) was comparable with the two-boxes solution, and was found to be highly successful in finding 100% functionally correct solutions. Especially when compared with single longer runs of the GA which, in many instances, was not able to find a correct solution at all.

The hope here is that that this technique may be applied to logic circuits - and especially those which have proved difficult to evolve e.g. the 3-bit multiplier. Results for these experiments will be reported in due course. Experiments have already begun to test the method described above in evolving analogue filter circuits using the PALMO system developed at Edinburgh University, with the preliminary results, which support the use of the technique, discussed in [11].

References

[1] Koza J. R., Andre D., Bennett III F. H., and Keane M. A., "Design of a High-Gain Operational Amplifier and Other Circuits by Means of Genetic Programming", in *Evolutionary Programming VI, Lecture Notes in Computer Science*, Vol. 1213, Springer-Verlag 1997, pp. 125 – 135.

[2] Thompson A., "An evolved circuit, intrinsic in silicon, entwined with physics", in Higuchi T., Iwata M., and Liu W., (Editors), *Proceedings of The First International Conference on Evolvable Systems: From Biology to Hardware (ICES96), Lecture Notes in Computer Science*, Vol. 1259, Springer-Verlag, Heidelberg, 1997., pp. 390 – 405.

[3] Thompson A., "On the Automatic Design of Robust Electronics Through Artificial Evolution.", in Sipper, M., Mange, D. and Perez-Uribe, A. (Editors), *Proceedings of The Second International Conference on Evolvable Systems: From Biology to Hardware (ICES98), Lecture Notes in Computer Science*, Vol. 1478, Springer-Verlag, 1998, pp. 13 – 24.

[4] Higuchi T., Iwata M., Kajitani I., Iba H., Hirao Y., Furuya T., and Manderick B., "Evolvable Hardware and Its Applications to Pattern Recognition and Fault-Tolerant Systems", in *Lecture Notes in Computer Science - Towards Evolvable Hardware*, Vol. 1062, Springer-Verlag, 1996., pp. 118 – 135.

[5] Goeke M., Sipper M., Mange D., Stauffer A., Sanchez E., and Tomassini M., "Online Autonomous Evolware", in Higuchi T., Iwata M., and Liu W., (Editors), *Proceedings of The First International Conference on Evolvable Systems: From Biology to Hardware (ICES96), Lecture Notes in Computer Science*, Vol. 1259, Springer-Verlag, 1997, pp. 96 – 106.

[6] Miller J. F. and Thomson P., "Aspects of Digital Evolution: Geometry and Learning." in Sipper, M., Mange, D. and Perez-Uribe, A. (Editors), *Proceedings of The Second International Conference on Evolvable Systems: From Biology to Hardware (ICES98), Lecture Notes in Computer Science*, Vol. 1478, Springer-Verlag, 1998, pp. 25 – 35.

[7] Miller J. F., Thomson P., and Fogarty T. C., "Designing Electronic Circuits Using Evolutionary Algorithms. Arithmetic Circuits: A Case Study.", in *Genetic Algorithms and Evolution Strategies in Engineering and Computer Science*: D. Quagliarella, J. Periaux, C. Poloni and G. Winter (eds), Wiley, 1997.

[8] Miller J. F. and Thomson P., "Designing Routable Circuits for FPGAs using Artificial Evolution.", submitted to *IEE Proceedings on Computers and Digital Techniques – Special Issue on Reconfigurable Computing.*, 1999.

[9] Miller J. F., Job D. and Vassilev V. K., "Principles in the Evolution of Digital Circuits – Part I.", in *The Journal of Genetic Programming and Evolvable Machines.*, 2000, to appear.

[10] Koza J. R., *Genetic Programming*, The MIT Press, Cambridge, Mass., 1992.

[11] Hamilton A., Thomson P. and Tamplin M., "Experiments in Evolvable Filter Design using Pulse Based Programmable Analogue VLSI Models.", to appear in *ICES2000 – The Third International Conference on Evolvable Systems: From Biology to Hardware.*, 2000.

[12] Miller J. F. and Thomson P., "Cartesian Genetic Programming.", to appear in *EuroGP – International Conference on Genetic Programming.*, 2000.

Evolutionary Robots with Fast Adaptive Behavior in New Environments

Joseba Urzelai and Dario Floreano

Laboratory of Microprocessors and Interfaces (LAMI)
Swiss Federal Institute of Technology, CH-1015 Lausanne (EPFL)
Joseba.Urzelai@epfl.ch, Dario.Floreano@epfl.ch

Abstract. This paper is concerned with adaptation capabilities of evolved neural controllers. A method consisting of encoding a set of local adaptation rules that synapses obey while the robot freely moves in the environment [6] is compared to a standard fixed-weight network. In the experiments presented here, the performance of the robot is measured in environments that are different in significant ways from those used during evolution. The results show that evolutionary adaptive controllers can adapt to environmental changes that involve new sensory characteristics (including transfers from simulation to reality) and new spatial relationships.

1 Evolution and Adaptation

Evolutionary algorithms are widely used in autonomous robotics in order to solve a large variety of tasks in several kind of environments. However, evolved controllers become well adapted to environmental conditions used during evolution, but often do not perform well when conditions are changed. Under these circumstances, it is necessary to carry on the evolutionary process, but this might take long time.

Combination of evolution and learning has been shown to be a viable solution to this problem by providing richer adaptive dynamics [1] than in the case where parameters are entirely genetically-determined. A review of the work combining evolution and learning for sensory-motor controllers can be found in [5, 9].

Instead of simply combining off-the-shelf evolutionary and learning algorithms, in previous work we presented an approach capable of generating adaptive neural controllers by evolving a set of simple adaptation rules [6]. The method consists of encoding on the genotype a set of modification rules that perform Hebbian synaptic changes [2–4] through the whole individual's life. The results showed that evolution of adaptive individuals generated viable controllers in much less generations and that these individuals displayed more performant behaviors than genetically-determined individuals.

In this paper, we describe two new sets of experiments conceived to measure the *adaptation* capabilities of this approach in environments that are different from those used during evolution. The results are compared to standard evolution of synaptic weights and to evolution of noisy synaptic weights (control condition).

J. Miller et al. (Eds.): ICES 2000, LNCS 1801, pp. 241–251, 2000.

Fig. 1. A mobile robot Khepera gains fitness by finding as fast as possible the stick under the floor. The walls are covered with white paper and the floor is transparent. The robot has a sensor pointing downwards that can detect the stick when it passes over it. The stick can be positioned at any location under the floor.

The sources of change address two major aspects of behavioral robustness: sensory appearance and spatial relationships of key-features of the environment.

2   Experiment I: Changing Sensory Appearances

A mobile robot Khepera is positioned in the rectangular environment shown in figure 1. The walls are covered with paper and the floor, which is transparent, is placed on four supports. A stick is positioned at a random location under the floor[1]. Each individual of the population is tested on the same robot, one at a time, for a maximum of 500 sensory motor cycles, each cycle lasting 100 ms. At the beginning of an individual's life, the robot and the stick are positioned at random positions.

The fitness function selects individuals capable of finding the stick in the shortest time,

$\Phi = 1 - \frac{t}{500}$,

where t represents the number of sensory motor cycles spent by the robot before finding the stick. Since the robot is not allowed to be on the target at the initial cycle, the fitness will never be 1.0. A robot that cannot manage to find the target will be scored with 0.0 fitness.

A light sensor placed under the robot is used to detect the stick and compute the fitness, but it is not given as input to the controller. Once the robot has found the stick or 500 cycles have passed, the robot and the target are randomly repositioned.

Under these circumstances, the exploration strategies used by the robots will depend much on the sensory appearance of the walls. Therefore, in this

---

[1] A similar environment has been used in simulation by Nolfi [10] for different experimental purposes.

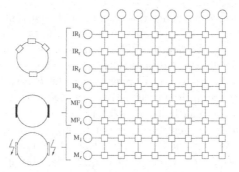

Fig. 2. The neural controller is a fully-recurrent discrete-time neural network composed of 8 neurons giving a total of 8 x 8= 64 synapses (here represented as small squares of the unfolded network). 6 sensory neurons receive additional input from the infrared sensors positioned around the body of the robot and from the motors (l=left; r=right; f=front; b=back). *IR*=Infrared Proximity sensors; *MF*=Motor feedback. Two motor neurons *M* do not receive sensory input; their activation sets the speed of the wheels ($M_i > 0.5$ forward rotation; $M_i < 0.5$ backward rotation)

experiment the sensory characteristics of the wall surfaces will be changed *after* evolution.

The controller is a fully-recurrent discrete-time neural network. Neurons are updated every 100 ms according to the following equation,

$$y_i \leftarrow \sigma \left(\sum_{j=0}^{N} w_{ij} y_j \right) + I_i,$$

where y_i is the activation of the ith neuron, w_{ij} is the strength of the synapse between presynaptic neuron j and postsynaptic neuron i, N is the number of neurons in the network, $0 \leq I_i < 1$ is the corresponding external sensory input, and $\sigma(x) = (1 + e^x)^{-1}$ is the sigmoidal activation function. $I_i = 0$ for the motor neurons. The controller has access to two types of sensory information: infrared light (object proximity) and speeds of the wheels (motor feedback). The active infrared sensors positioned around the robot detect proximity of walls (up to 4 cm). Their values are pooled into four pairs and the average reading of each pair is passed to a corresponding neuron. Rotation speeds of the wheels[2] are normalized and passed to the corresponding feedback neurons. An additional neuron is used as bias in order to excite the network when it does not receive any sensory input. Two output neurons are used to set the rotation speeds of the wheels.

Each synaptic weight w_{ij} can be updated after every sensory-motor cycle (100 ms) using one of the four modification rules specified in the genotype.[3] The four rules are called Hebbian because they are a function of the pre-synaptic ac-

---

[2] Rotation speeds of the wheels can be different from the values set by the motor neurons. For example, when the robot pushes against a wall, motor neurons may output forward rotations but real rotations are 0.

[3] These four rules co-exist within the same network.

tivation, of the post-synaptic activation, and of the current value of the weight itself. The *Plain Hebb rule* strengthens the synapse proportionally to the correlated activity of the two neurons. The *Postsynaptic rule* behaves as the plain Hebb rule, but in addition it weakens the synapse when the postsynaptic node is active but the presynaptic is not. Conversely, in the *Presynaptic rule* weakening occurs when the presynaptic unit is active but the postsynaptic is not. Finally, the *Covariance rule* strengthens the synapse whenever the difference between the activations of the two neurons is less than half their maximum activity, otherwise the synapse is weakened. Synaptic strength is maintained within a range [0, 1] (notice that a synapse cannot change sign) by adding to the modification rules a self-limiting component inversely proportional to the synaptic strength itself [2, 3, for more details].

Two types of genetic (binary) encoding are considered: *Synapse Encoding* and *Node Encoding*. Synapse Encoding is also known as direct encoding [11]. Every synapse is individually coded on 5 bits, the first bit representing its sign and the remaining four bits its properties (either the weight strength or its adaptive rule). Node Encoding instead codes only the properties of the nodes of the network. These properties are then applied to all its incoming synapses (consequently, all incoming synapses to a given node have the *same* properties). Each node is characterized by 5 bits, the first bit representing its sign and the remaining four bits the properties of its incoming synapses. Synapse Encoding allows a detailed definition of the controller, but for a fully connected network of N neurons the genetic length is proportional to N^2. Instead Node Encoding requires a much shorter genetic length (proportional to N), but it allows only a rough definition of the controller.

Independently of the type of genetic encoding, the following three types of properties can be encoded on the last 4 bits. A) *Genetically determined*: Weight strength. The synaptic strength is genetically determined and cannot be modified during "life". B) *Adaptive synapses*: Adaptive rule on 2 bits (four rules) and learning rate (0.0, 0.3, 0.6, 0.9) on the remaining 2 bits. The synapses are always randomly initialized when an individual starts its life and then are free to change according to the selected rule. C) *Noisy synapses*: Weight strength on 2 bits and a noise range on the remaining two bits (0.0, ±0.3, ±0.6, ±0.9). The synaptic strength is genetically determined at birth, but a random value extracted from the noise range is freshly computed and added after each sensory motor cycle. A limiting mechanism cuts off sums that exceed the synaptic range [0, 1]. This latter condition is used as a control condition to check whether the effects of Hebbian adaptation amount to random synaptic variability [6, for more details].

2.1   Results

An initial set of experiments has been carried out in simulations sampling sensor activations separately for white, gray, and black walls and adding 5% uniform noise to these values [8]. Robots are evolved in environments with white walls. For each experimental condition (adaptive, genetically-determined, noisy), 10

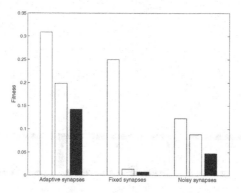

Fig. 3. Comparison of adaptive synapses with Node Encoding (*left*) versus genetically-determined synapses with Synapse Encoding (*center*) and noisy synapses with Node Encoding (*right*) in white, gray, and black environments.

different[4] populations of 100 individuals each have been independently evolved for 200 generations. Each individual is tested three times and the fitness value is averaged. The 20 best individuals reproduce by making 5 copies of their genetic string. Strings are crossed over with probability 0.2 and mutated with probability 0.05 (per bit). In the case of adaptive synapses, synaptic weights of individuals are randomly initialized within the range [0.0, 0.1] at the beginning of each test.

Since we are interested in adaptation capabilities of evolved individuals, after 200 generations evolution is stopped and the best individual of the last generation for each of the 10 populations is tested 10 times in the original environment (white walls), 10 times in an environment where walls have been covered with gray paper, and 10 times in an environment with black walls.

Figure 3 shows average fitnesses corresponding to environments with white, gray, and black walls in the case of individuals with adaptive synapses and Node Encoding (left), individuals with genetically-determined synapses and Synapse Encoding[5] (center), and individuals with noisy synapses and Node Encoding (right). Although performance decreases in gray and black environments, adaptive individuals are capable of successfully finding the target area in all conditions. Instead, genetically-determined individuals can find the target area only in the environment that has been used during evolution (white walls). When tested in gray and black environments, only a few lucky individuals that encounter the target before a wall have non-zero fitness values. Individuals with noisy synapses score very low fitness values in all conditions but they generalize better than genetically-determined individuals[6].

---

[4] Using different sequences of random number.

[5] Node Encoding for fixed synapses was not capable of solving the original problem, therefore we report results for Synapse Encoding.

[6] Notice that adaptive individuals report better fitness also in the evolutionary environment. The performance issue has been addressed in another paper [6].

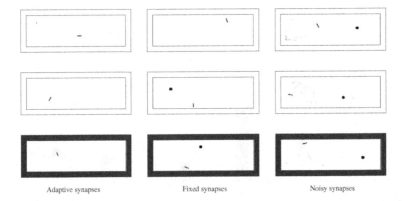

Adaptive synapses   Fixed synapses   Noisy synapses

Fig. 4. Behaviors of individuals with adaptive synapses (left), genetically-determined synapses (center), and noisy synapses (right) in environments with white (up), gray (center), and black (bottom) walls. Individuals belong to the last generation.

Figure 4 displays the behaviors of individuals with adaptive synapses (left), genetically-determined synapses (center), and noisy synapses (right) in environments with white (up), gray (center), and black (bottom) walls. The behavior of the adaptive individual is not considerably affected by the color of the walls and it reaches the target area in all conditions. Instead, the genetically-determined individual can reach the target area only when walls are covered with white paper, but gets stuck on gray and black walls. Since darker walls are detected only when the robot gets closer, a behavioral strategy successful for white walls can cause collisions for dark walls. The individual with noisy synapses takes advantage of the random variability to get away from the walls but it scores a low performance because its strategy is based in a local random search. The robot eventually manages to reach the target when its initial position is relatively close to the stick but it fails when it is far away.

2.2 From Simulations to Real Robots

Another way of measuring the adaptive abilities of evolved controllers is to transfer them from simulated to real robots. Since physical robots and environments inevitably have characteristics different from simulations, solutions evolved in simulation typically fail when tested on real robots.

The solutions envisaged so far consist of incorporating special types of noise tailored to sensory-motor properties of the robot [8], or to vary key-features of the environment during simulated evolution [7]. The success of both methods depends upon the ability of the experimenter to spot crucial aspects of variation that must be considered in the simulations. Another solution consists of carrying on artificial evolution in the new conditions [3], but this can take long time.

In another set of experiments, we have tested the best individuals of the last generation for each of the 10 populations evolved in simulation on a real

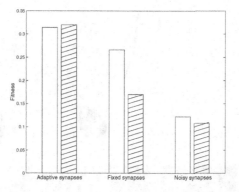

Fig. 5. Comparison of generalization capabilities in the transfer from simulations to real robots in the case of adaptive synapses with Node Encoding (*left*) versus genetically-determined synapses with Synapse Encoding (*center*) and noisy synapses with Node Encoding (*right*). White bars correspond to the performance in the white simulated environment and striped bars represent the performance on real robots in a white environment.

Khepera robot positioned in an environment where walls are covered with white paper (figure 1). Each individual is tested 3 times and the fitness is averaged over. Figure 5 shows that the performance of adaptive individuals is not affected by the transfer to the physical environment, whereas genetically-determined individuals report a significative fitness loss. Individuals with noisy synapses are not affected by the transfer because their behavior is always random and not effective in both simulated and physical environments. A major reason for failure of genetically-determined individuals is that their spiralling strategy often results in rotation without displacement probably caused by the new sensory and motor responses of the real robot. The same pattern of results holds for tests in gray and black physical environments (data not shown).

3   Experiment II: Changing Spatial Relationships

Whereas the experiments described above were conceived to address mostly variation induced by new sensory responses, in this section we address variation induced by changed spatial relationships. To this end, we resort to an experimental situation where behavioral success is linked to the ability to relate different parts of the environment [6].

A mobile robot Khepera equipped with a vision module is positioned in the rectangular environment shown in figure 6. A light bulb is attached on one side of the environment. This light is normally off, but it can be switched on when the robot passes over a black-painted area on the opposite side of the environment. A black stripe is painted on the wall over the light-switch area. Each individual of the population is tested on the same robot, one at a time, for 500 sensory

Fig. 6. A mobile robot equipped with a vision module gains fitness by staying on the gray area only when the light is on. The light is normally off, but it can be switched on if the robot passes over the black area positioned on the other side of the arena. The robot can detect ambient light and the color of the wall, but not the color of the floor.

motor cycles, each cycle lasting 100 ms. At the beginning of an individual's life, the robot is positioned at a random position and orientation and the light is off.

The fitness function is described as the number of sensory motor cycles spent by the robot on the gray area beneath the light bulb *when the light is on* divided by the total number of cycles available (500). In order to maximize this fitness function, the robot should find the light-switch area, go there in order to switch the light on, and then move towards the light as soon as possible, and stand on the gray area[7]. Since this sequence of actions takes time (several sensory motor cycles), the fitness of a robot is never 1.0. Also, a robot that cannot manage to complete the entire sequence is scored with 0.0 fitness.

A light sensor placed under the robot is used to detect the color of the floor— white, gray, or black— and passed to a host computer in order to switch on the light bulb and compute fitness values. The output of this sensor is *not* given as input to the neural controller. After 500 sensory motor cycles, the light is switched off and the robot is repositioned by applying random speeds to the wheels for 5 seconds.

In a previous article we showed that evolution of adaptive synapses provides a number of advantages with respect to evolution of synaptic weights for this behavioral task. It can generate viable controllers in much less generations and evolved controllers display more performant behaviors [6].

Here we describe a new set of experiments, where the best individuals of the last generation are tested in environments where the light-switching area,

---

[7] Notice that the fitness function does not explicitly reward this sequence of actions, but only the final outcome of the overall behavior chosen by the robot. Therefore, we call it a *behavior-based* fitness function.

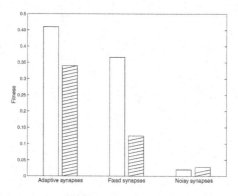

Fig. 7. Comparison of adaptive synapses with Node Encoding (*left*) versus genetically-determined synapses with Synapse Encoding (*center*) and noisy synapses with Node Encoding (*right*) in the environment used during evolution (white) and in an environment where light-switching area and fitness area are randomly positioned (striped).

the fitness area, and the robot are located at random positions at the beginning of each individual's life. Since in the original experiment the positions of the light-switching area and the fitness area were constant for every individual, this experiment gives us a measure of adaptation capabilities of evolved individuals to new spatial relationships. In order to automate the re-arrangement of the environment, these experiments have been carried out in simulation. The best individuals for each of the 10 populations evolved in the environment of figure 6 are tested in 3 new environments with different random spatial relationships.

The results reported in figure 7 show that individuals with adaptive synapses are much more robust to new configurations of the environment than individuals with genetically-determined synapses. Average performance loss is 25% in the case of adaptive individuals (left), but is about 65% in the case of genetically-determined individuals (center). Individuals with noisy synapses (right) score very low fitness in both cases.

The fact that genetically-determined individuals performed very poorly in new environments indicates that the solutions generated by evolution here are tightly coupled to the geometry and the disposition of the environment. Evolution shapes the individuals in order to take advantage of specific environmental aspects, such as the size of the arena and the position of the light-switching area and of the fitness area. Instead, evolution of adaptive synapses is capable of generating more general solutions that produce performant behavior for a large variety of environmental dispositions. This is shown in figure 8: a genetically-determined individual (center) is capable of solving the task in the original environment by performing circular movements and avoiding the walls until it reaches the fitness area. However, these circular movements are not effective to approach the fitness area in the new environmental disposition. Instead, an adaptive individual (left) that is capable of solving the task in the original envi-

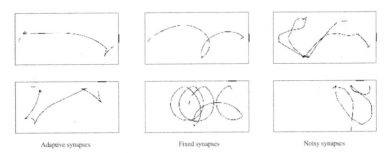

Fig. 8. Behaviors of individuals with adaptive synapses (left), genetically-determined synapses (center), and noisy synapses (right) in the original environment used during evolution (top) and in an environment where light-switching area and fitness area are positioned at new locations (bottom). The trajectory line is thin when the light is off and becomes thick when the light is turned on. Individuals are the best of the last generation evolved in the environment of top row.

ronment changes the strategy by performing some additional manoeuvres that allow the robot to reach the fitness area in the new environment. Individuals with noisy synapses (right) perform random trajectories in both cases.

4   Conclusions

We have shown through a set of systematic comparisons that evolution of adaptive synapses provides better adaptation capabilities than standard evolution of synaptic weights. Adaptive individuals are capable of successfully performing in environments that are different from the one used during evolution by adapting their strategy to the new constraints of the environment. Instead, genetically-determined individuals often fail in adapting to different environments because their behavior is tightly coupled to the characteristics of the environment used during evolution.

We have studied adaptation to two major sources of environmental change: new sensory appearances and new spatial relationships. In both cases, evolved adaptive controllers can autonomously modify their parameters and behavior online without requiring additional evolutionary training or ad-hoc manipulation of the evolutionary procedure. The control experiments with noisy synapses where changes are induced by random numbers rather than by genetically-determined rules, indicate that evolved adaptive networks modify their parameters in ways that are functionally related to the survival criterion. We are currently analyzing how these evolved controllers work and try to explain how they can adapt to new environmental conditions.

We have also shown that this approach is effective for the transfer of evolved controllers from simulations to real robots. Keeping in mind this idea, one of our current projects aims at studying the applicability of this approach to cross-platform evolution [3]. The experiment will consist in transferring the controller evolved for the Khepera robot to a bigger Koala robot and in testing its performance in a scaled-up version of the light-switching environment.

Acknowledgements

Joseba Urzelai is supported by grant nr. BF197.136-AK from the Basque government.

References

1. R. K Belew and M. Mitchell, editors. *Adaptive Individuals in Evolving Populations. Models and Algorithms.* Addison-Wesley, Redwood City, CA, 1996.
2. D. Floreano and F. Mondada. Evolution of plastic neurocontrollers for situated agents. In P. Maes, M. Mataric, J-A. Meyer, J. Pollack, H. Roitblat, and S. Wilson, editors, *From Animals to Animats IV: Proceedings of the Fourth International Conference on Simulation of Adaptive Behavior*, pages 402–410. MIT Press-Bradford Books, Cambridge, MA, 1996.
3. D. Floreano and F. Mondada. Evolutionary neurocontrollers for autonomous mobile robots. *Neural Networks*, 11:1461–1478, 1998.
4. D. Floreano and S. Nolfi. Adaptive behavior in competing co-evolving species. In P. Husbands and I. Harvey, editors, *Proceedings of the 4th European Conference on Artificial Life*, Cambridge, MA, 1997. MIT Press.
5. D. Floreano and J. Urzelai. Evolution and learning in autonomous robots. In T. Mange and M. Tomassini, editors, *Bio-inspired Computing Systems*. PPUR, Lausanne, Switzerland, 1998.
6. D. Floreano and J. Urzelai. Evolution of Neural Controllers with Adaptive Synapses and Compact Genetic Encoding. In D. Floreano, J.D. Nicoud, and F. Mondada, editors, *Advances In Artificial Life: Proceedings of the 5th European Conference on Artificial Life (ECAL'99)*, pages 183–194. Springer Verlag, Berlin, 1999.
7. N. Jakobi. Half-baked, ad-hoc and noisy: Minimal simulations for evolutionary robotics. In P. Husbands and I. Harvey, editors, *Advances in Artificial Life: Proceedings of the 4th European Conference on Artificial Life*, pages 348–357. MIT Press, 1997.
8. O. Miglino, H. H. Lund, and S. Nolfi. Evolving Mobile Robots in Simulated and Real Environments. *Artificial Life*, 2:417–434, 1996.
9. S. Nolfi and D. Floreano. Learning and evolution. *Autonomous Robots*, 7(1):forthcoming, 1999.
10. S. Nolfi and D. Parisi. Learning to adapt to changing environments in evolving neural networks. *Adaptive Behavior*, 5:75–98, 1997.
11. X. Yao. A review of evolutionary artificial neural networks. *International Journal of Intelligent Systems*, 4:203–222, 1993.

The Advantages of Landscape Neutrality in Digital Circuit Evolution

Vesselin K. Vassilev[1] and Julian F. Miller[2]

[1] School of Computing, Napier University
Edinburgh, EH14 1DJ, UK
v.vassilev@dcs.napier.ac.uk
[2] School of Computer Science, University of Birmingham
Birmingham, B15 2TT, UK
j.miller@cs.bham.ac.uk

Abstract. The paper studies the role of neutrality in the fitness landscapes associated with the evolutionary design of digital circuits and particularly the three-bit binary multiplier. For the purpose of the study, digital circuits are evolved extrinsically on an array of logic cells. To evolve on an array of cells, a genotype-phenotype mapping has been devised by which neutrality can be embedded in the resulting fitness landscape. It is argued that landscape neutrality is beneficial for digital circuit evolution.

1 Introduction

Digital circuit evolution is a process of evolving configurations of logic gates for some prespecified computational program. Often the aim is for a highly efficient electronic circuit to emerge in a population of instances of the program. Digital electronic circuits have been evolved intrinsically [1] and extrinsically [2–6]. The former is associated with an evolutionary process in which each evolved electronic circuit is built and tested on hardware, while the latter refers to circuit evolution implemented entirely in software using computer simulations.

A possible way to study the evolvability of digital circuits is to consider the evolutionary design as a search on a *fitness landscape* [7]. The metaphor comes from biology to represent adaptive evolution as a population flow on a mountainous surface in which the elevation of a point qualifies how well the corresponding organism is adapted to its environment [8]. In evolutionary computation the fitness landscapes are simply search spaces that originate from the combination of the following objects

1. A set of configurations that are often referred to as *genotypes*.
2. A cost function that evaluates the configurations, known in evolutionary computation as a *fitness function*.
3. A topological structure that allows relations within the set of configurations.

These define the structure of the fitness landscape. Recently it has been shown that the landscape structure affects the evolvability of a variety of complex

J. Miller et al. (Eds.): ICES 2000, LNCS 1801, pp. 252–263, 2000.
© Springer-Verlag Berlin Heidelberg 2000

systems [9–15]. In evolutionary computation the notion of evolvability refers to the efficiency of evolutionary search.

The circuit evolution landscapes associated with the evolution of various arithmetic functions were studied in [7, 16, 17]. It was shown that these landscapes are products of three subspaces with different landscape characteristics. These are the functionality, internal connectivity, and output connectivity landscapes. In general they are characterised with neutral networks and sharply differentiated plateaus. A set of genotypes defines a *neutral network* if the set represents a connected subgraph of genotypes with equal fitnesses [18, 19]. This characteristic of fitness landscapes is referred to as *neutrality*.

The landscape neutrality in digital circuit evolution originates mainly from the genotype-phenotype mapping by which a digital circuit is encoded into a genotype. The mapping is defined in such a way that it allows neutrality. This affords a study of the important question of the role of landscape neutrality in the evolutionary design of digital circuits. For the purpose of this study, digital circuits are evolved extrinsically. This allows freedom to explore the methodology and thus to extract principles of the evolutionary design of circuitry in general [17].

Studies in evolutionary biology suggest that adaptive evolution is facilitated by a genetic variation that is due to neutral or nearly neutral mutations [20–23]. In [24] it was suggested that the role of landscape neutrality for adaptive evolution is to provide a "path" for crossing landscape regions with poor fitness. This implies a scenario of adaptive evolution in which a population evolves on a neutral network until another neutral network with a higher fitness is reached [25]. Similar conclusions were attained in a study of the technique of genotype-phenotype mapping that appeared to be suitable for solving constrained optimisation problems by genetic programming [26]. Evidence that fitness improvements may occur in a genetically converged population due to neutrality was given also in [27].

Does the evolutionary design of digital circuits benefit from the neutrality in the fitness landscapes? In this paper this question is answered in the affirmative. The paper is organised as follows. The next section introduces digital circuit evolution in greater details. Section 3 represents the evolution of a three-bit multiplier. The neutrality of the fitness landscapes is studied in section 4. Section 5 studies the benefit of landscape neutrality for the adaptive design of digital circuits. The paper closes with a summary and suggestions for future work.

2   Digital Circuit Evolution

The technique used in the evolutionary design of digital circuits in this paper is that adopted in the framework of Cartesian Genetic Programming [28, 29] and it uses an evolutionary algorithm with truncation selection and mutation. The latter is defined as a percentage of the genes in a single genotype which are to be randomly mutated. In this paper the percentage chosen results in 3 mutated genes per genotype. The algorithm deals with a population of digital feed-forward electronic circuits that are instances of a particular program. The population consists of $1 + \lambda$ genotypes where λ is usually about 4. Initially the elements of the population are chosen at random. The fitness value of each genotype is evaluated, by calculating the number of total correct outputs of the encoded electronic circuit in response to all appropriate input combinations. For convenience, in this paper the fitness values are scaled in the interval $[0, 1]$. To update the population, the mutation operator is applied to the fittest genotype, to generate offspring. These together with the parent constitute the new population. This mechanism of population update has some similarities with that employed in other evolutionary techniques such as $(1+\lambda)$ Evolution Strategy [30, 31] and the Breeder Genetic Algorithm [32].

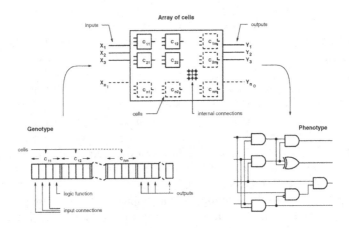

Fig. 1. The phenotype that is a digital circuit is encoded within a genotype by an array of logic cells.

To encode a digital electronic circuit into a genotype, a genotype-phenotype mapping is defined. This is done via rectangular array of cells each of which is an atomic two-input logic gate or a multiplexer. Thus the genotype is a linear string of integers and it consists of two different types of genes that are responsible for the functionality and the routing of the evolved array. The genotype is characterised by four parameters of the array of cells: the *number of allowed logic functions*, the *number of rows*, the *number of columns*, and *levels-back*. The first parameter defines the functionality of logic cells, while the latter three parameters determine the layout and routing of the array. Note that the number of inputs and outputs of the array are specified by the objective function. The genotype-phenotype mapping is defined using an array of $n \times m$ three-input cells with n_I inputs and n_O outputs, and together with the genotype representation is shown in Figure 1. The array is a composition of cells each of which can be any allowed two-input logic gate or alternatively a multiplexer.

The internal connectivity of the array is defined by the connections between the array cells. The inputs of each cell are only allowed to be inputs of the array or outputs of the cells with lower column numbers. The internal connectivity is also dependent on the levels-back parameter that defines the array inputs and cells to which a cell or an array output can be connected in the following manner. Consider that the levels-back parameter is equal to L. Then cells can be connected to cells from L preceding columns. If the number of preceding columns of a cell is less than L then the cell can also be connected to the inputs of the array. In this paper the array cells and outputs are maximally connectable since the number of rows is set to one and the levels-back is equal to the number of columns.

The gate array output connectivity is defined in a similar way. The output connections of the array are allowed to be outputs of cells or array inputs. Again, this is dependent on the neighbourhood defined by the levels-back parameter.

The genotype is a string of integers that encode either logic functions or connections. The logic functions are represented by letters associated with the allowed cell functionality. The connections are defined by indexes that are assigned to all inputs and cells of the array. Each array input X_k is labelled with $k-1$ for $1 \leq k \leq n_I$, and each cell c_{ij} is labelled with an integer given by $n_I + n(j-1) + i - 1$ for $1 \leq i \leq n$ and $1 \leq j \leq m$. Thus the genotype consists of groups of four integers that encode the cells of the array, followed by a sequence of integers that represent the indexes of the cells connected to the outputs of the array. The first three values of each group are the indexes of the cells to which the inputs of the encoded cell are connected. If the cell represents a two-input logic function, then the third connection is redundant. This type of redundancy is referred to as *input redundancy*. The last integer of the group represents the logic function of the cell. Cells may also not have their outputs connected in the operating circuit. This is another form of redundancy called *cell redundancy*. The redundancy in the genotype related to the function of the array may also be *functional redundancy*. This is the case in which the number of cells of a digital circuit is higher than the optimal number needed to implement this circuit.

3   Evolution of a Three-bit Multiplier

To study the role of landscape neutrality in the evolutionary design of digital circuits, a three-bit multiplier is evolved using binary multiplexers. The three-bit multiplier is a good candidate to be used in this study for the following reasons: firstly, the circuit is difficult to evolve, and secondly, it is a fundamental building block of many digital devices. In addition it has been shown that the fitness landscapes associated with the evolution of the three-bit multiplier are similar to the landscapes of other arithmetic functions in terms of landscape neutrality [17]. The binary multiplexers are defined by the universal-logic function

$$f(a,b,c) = a \cdot \bar{c} + b \cdot c \tag{1}$$

taken four times with inputs a and b inverted in various ways (gates $16 - 19$ as labelled in [17]). The reason for using only multiplexers is to simplify the evolutionary model by allowing only the existence of cell redundancy and functional redundancy. 100 evolutionary runs of 10 million generations were carried out. The array had 1×24 cells and the levels-back was set to 24. 27 perfect solutions were found, three of which were circuits that required 21 gates. For each evolutionary run the best fitness of the population and the corresponding genotype were recorded for the generations in which improvements of the fitness have been attained. In addition, the number of neutral changes between every two fitness improvements was evaluated. Thus the number of neutral mutations for each fitness improvement were calculated cumulatively. The aim is to attain understanding of the process of evolving digital circuits, particularly the three-bit multiplier. The results for a typical evolutionary run in which a three-bit multiplier of 21 gates was obtained are given in Figure 2. The figure represents (a) the best fitness, and (b) the cumulative number of neutral mutations in the (1) functionality, (2) internal connectivity, and (3) output connectivity configurations. The circuit was attained at generation $4,970,271$, and its schematic is given in Figure 3. The circuit is efficient in term of gate usage, since it consists of 21 logic gates that is 20% less than the number of gates of the best conventional design. The logic gates used in the figure are multiplexers (given with rectangles), AND, OR, and XOR. The two-input gates in the depicted multiplier came about because some multiplexers had two inputs connected together.

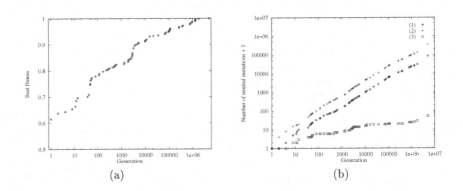

(a)   (b)

Fig. 2. The evolution of a three-bit multiplier: (a) the best fitness, and (b) the cumulative number of neutral mutations in the (1) functionality, (2) internal connectivity, and (3) output connectivity configurations, recorded at each fitness increase (74 in total).

Figure 2a shows that the best fitness is marked by periods of a sharp fitness increase followed by periods of stasis or a slight fitness increase. It is hypothesised that each period of stasis is a hidden process of search performed by neutral walks so that the population could traverse wide search space areas with lower or equal fitness. To investigate this phenomenon the number of all neutral mutations at each generation for functionality, input connectivity and output connectivity configurations are also evaluated cumulatively (Figures 2b). It is

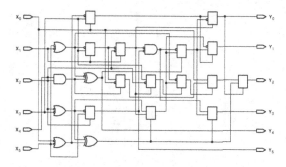

Fig. 3. The schematic of the evolved three-bit multiplier. The most significant bits are inputs X_0 and X_3, and output Y_0.

shown that the cumulative number of neutral mutations increases with a lower than linear rate at each generation, and the evolutionary process is constantly accompanied by neutral mutations. The plots also show that neutral changes are most likely to appear in the input connectivity configurations, less in the functionality configurations, and least in the output connectivity configurations.

4   Landscape Neutrality

The landscapes of the evolved three-bit multiplier result from the Cartesian product of three configuration spaces [7] defined on the functionality and connectivity alphabets that in this particular case have sizes $l_\alpha = 4$ and $l_\beta = 24$, respectively. The former alphabet is defined by the number of allowed logic functions while the latter is defined by the levels-back parameter. The structure of the three subspaces - functionality, internal connectivity, and output connectivity - did not differ significantly from the structure of the landscapes associated with other electronic circuits, such as two-bit and three-bit multipliers evolved for various values of the functionality and connectivity parameters [17]. They are characterised with neutrality that prevails over the landscape smoothness and ruggedness in the internal connectivity subspace. This was not valid for the functionality and output connectivity landscapes. It was also found that the neutrality of the output connectivity landscapes was more strongly dominated by the landscape ruggedness than that found in the functionality landscapes. These findings were revealed by studying the information characteristics [33] of time series obtained via random walks with respect the three subspaces.

An interesting issue related to the role of neutrality in the evolution of digital circuits is the relation between the size and the height of the landscape plateaus. It is believed that the neutral walks are longer at a lower altitude fitness level. Thus it can be surmised that the length of the neutral walks will decrease as the best fitness increases. The reason is that the genotype redundancy is expected to decrease in an efficient evolutionary search. This can be illustrated by measuring the length of the neutral walks that start from those genotypes recorded at

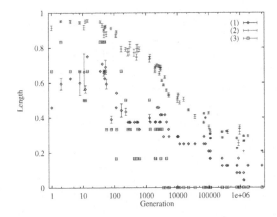

Fig. 4. The length of neutral walks on (1) functionality, (2) internal connectivity, and (3) output connectivity landscapes scaled in the interval [0, 1].

every fitness increase in the evolutionary run shown in Figure 2. The algorithm of neutral walks as given in [18] is defined as follows: start from a configuration, generate all neighbours, select a neutral one at random that results in an increase in the distance from the starting point and continue moving until the distance cannot be further increased. 1,000 neutral walks per configuration were performed. The means and the standard deviations of the lengths scaled in the interval [0, 1] are given in Figure 4. The scaling was done by dividing the length of each neutral walk by the length of the corresponding configuration. Note that the functionality, internal connectivity, and output connectivity configurations consist of 24, 72, and 6 genes, respectively. The figure confirms the findings of the information analysis of these landscapes. It is also shown that the length of the neutral walks decrease during the evolution with a higher than linear rate. An interesting result related to the evolved three-bit multiplier is that the length of the neutral walks on internal connectivity subspaces that start from the obtained functionally correct digital circuit exceeds the expected length regarding the cell redundancy. For this genotype, the expected length of the internal connectivity configuration is 9 since the number of redundant cells is 3 (9 redundant genes). However, the measured lengths were about 15. Therefore, there exist functional redundancy in the evolved internal connectivity configuration. This implies that some of the multiplexers of the circuit might be replaced with two-input logic gates this is also revealed by Figure 3.

5   Neutral Mutations and Search

The results represented thus far showed that digital circuit evolution is accompanied by a random genetic drift caused by neutral mutations. These are more likely to appear during the search on the functionality and input connectivity

landscapes since these subspaces are characterised with neutral networks that
originate from the cell redundancy. The amount of neutral changes in the output
connectivity configuration is much lower than in the functionality and internal
connectivity ones. Note that the neutrality of the output connectivity subspace
is determined only by the functional redundancy. It was also revealed that the
cumulative number of neutral changes in the functionality and internal connectivity configurations increases during the evolutionary run with a lower than
linear rate. This is to be expected since the fitness increase during the evolutionary run reduces the redundancy in general. This process in itself affects the
landscape neutrality so that the size of neutral areas decreases with a higher
than linear rate. The decrease of the landscape neutrality was revealed by measuring the length of the neutral walks that start from those genotypes recorded
at every fitness increase (section 4). The interesting question here is how the
increase of the cumulative number of neutral changes relates to the decrease of
the landscape neutrality during the evolution. This is answered by Figure 5 that
shows the derivatives (absolute values) of the plots of (1) the cumulative number
of neutral changes (Figure 2b), and (2) the length of neutral walks (Figure 4).
The derivatives are calculated for each generation characterised with a fitness

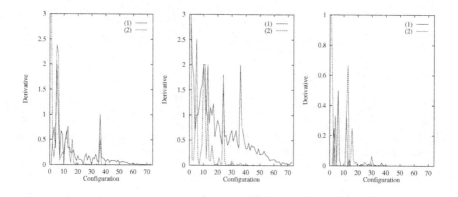

Fig. 5. Derivatives of the plots depicted in (1) Figure 2b and (2) Figure 4 calculated at
each generation characterised with a fitness increase (74 in total): (left) functionality,
(centre) internal connectivity, and (right) output connectivity configurations.

increase. The figure shows that the number of neutral changes decreases more
slowly than the length of the neutral walks. This also holds for the output connectivity subspaces, although it is difficult to see this in the figure. The findings
suggest that in the beginning of the run, neutral changes occur as a consequence
of the high redundancy in the genotype. This however does not appear as a
reason for the neutral changes at end of the run, since the redundancy becomes
low. It appears that the selective mechanisms promote the neutral changes since
this is the only feasible way for the population to explore the search space. This
is also indicated in the plot in Figure 6. The plot shows the Hamming distance

of every two consecutive genotypes each of which resulted in a fitness increase for the evolutionary run studied in section 3. The Hamming distance increases with the length of the periods of stasis that is another indication of the genetic drift caused by neutral changes. For instance, the Hamming distance of the last two genotypes obtained at generations $1,771,234$ and $4,970,271$ is equal to 75, although the difference between the fitness values of these genotypes is approximately 0.0027 (this is exactly a difference of one bit of the corresponding truth tables). This is a significant difference when considering that the length of the genotype is 102. The drift was attained after $281,163$ neutral mutations. Hence neutral evolution is vitally important for the search especially when close to the global maximum where the likelihood of deleterious mutations to occur is high.

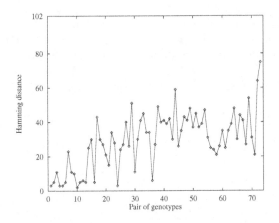

Fig. 6. The Hamming distance of two consecutive genotypes each of which represents a fitness increase in the evolutionary run shown in Figure 2.

The existence of neutrality helps the evolutionary design of digital circuits. Indeed if digital circuit evolution is implemented without neutral mutations, the result is not encouraging. This is illustrated by Figure 7. The figure shows the best fitness attained in the evolution of a three-bit multiplier in the experiment of 100 runs with "allowed" neutral mutations described in section 3, and the best fitness attained for 100 runs with "forbidden" neutral mutations. To allow neutral mutations, the algorithm was set up to choose a new parent even if the new fittest members of the population have fitness values that are equal to the fitness of the previous parent. Alternatively, to forbid neutral mutations, the algorithm was set up to change the parent only if a fitter member of the population occurs. The runs in which neutral mutations were allowed generated 27 perfect solutions: 5 with 24, 10 with 23, 9 with 22, and 3 with 21 logic gates. Although, the attained best fitness in the experiment with forbidden neutral mutations is fairly high, no perfect solution was evolved. This again supports the importance of landscape neutrality for the success of the evolutionary search.

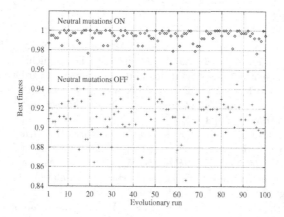

Fig. 7. The best fitness values of 2 × 100 evolutionary runs of 10 million generations with allowed (diamonds) and forbidden (crosses) neutral mutations.

The neutral evolution allows the population to avoid being trapped in a local optimum and thus to escape, and furthermore to cross wide landscape areas with lower fitness.

6   Summary

The importance of landscape neutrality for the evolution of digital circuits, particularly the three-bit multiplier, was revealed in a comparison between the amount of neutral changes and the size of the neutral areas during a successful evolutionary run. It was shown that the evolutionary process is accompanied by neutral mutations, the number of which, decreases with a lower rate when comparing with the decrease of the size of the neutral areas (Figure 5). Consequently, the neutral changes were employed in the evolutionary search (see also Figure 6). The landscape neutrality appeared to be vitally important for the evolutionary design of digital circuits in that it firstly prevents the evolved sub-circuit from deleterious mutations, and secondly, it allows the evolutionary search to avoid entrapment at local optima. This was empirically demonstrated in section 5 where it was shown that the search with allowed neutral changes is better.

Further research should be carried out to answer the following questions. How exactly does the circuit change in every period of stasis? How does evolution preserve the attained circuit modules? This remains for the future.

Acknowledgement

The authors would like to thank the anonymous reviewers for their valuable and detailed comments on the earlier draft of the paper.

References

1. Kajitani, I., Hushino, T., Nishikawa, D., Yokoi, H., Nakaya, S., Yamauchi, T., Inuo, T., Kajihara, N., Iwata, M., Keymeulen, D., Higuchi, T.: A gate-level ehw chip: Implementing GA operations and reconfigurable hardware on a single LSI. In Sipper, M., Mange, D., Pérez-Uribe, A., eds.: Proceedings of the 2nd International Conference on Evolvable Systems: From Biology to Hardware. Heidelberg, Springer-Verlag (1998) 1–12.
2. Higuchi, T., Niwa, T., Tanaka, T., Iba, H., de Garis, H., Furuya, T.: Evolving hardware with genetic learning: A first step towards building a Darwin machine. In Meyer, J.A., Roitblat, H.L., Stewart, W., eds.: From Animals to Animats II: Proceedings of the 2nd International Conference on Simulation of Adaptive Behaviour. Cambridge, MA, MIT Press (1993) 417–424.
3. Hemmi, H., Hikage, T., Shimohara, K.: Adam: A hardware evolutionary system. In: Proceedings of the 1st International Conference on Evolutionary Computation. Piscataway, NJ, IEEE press (1994) 193–196.
4. Hemmi, H., Mizoguchi, J., Shimohara, K.: Development and evolution of hardware behaviours. In Brooks, R., Maes, P., eds.: Artificial Life IV: Proceedings of the 4th International Workshop on the Synthesis and Simulation of Living Systems. Cambridge, MA, MIT Press (1994) 371–376.
5. Miller, J.F., Thomson, P., Fogarty, T.: Designing electronic circuits using evolutionary algorithms. arithmetic circuits: A case study. In Quagliarella, D., Periaux, J., Poloni, C., Winter, G., eds.: Genetic Algorithms and Evolution Strategies in Engineering and Computer Science. Wiley, Chechester, UK (1997) 105–131.
6. Iba, H., Iwata, M., Higuchi, T.: Machine learning approach to gate-level evolvable hardware. In Higuchi, T., Iwata, M., eds.: Proceedings of the 1st International Conference on Evolvable Systems: From Biology to Hardware. Heidelberg, Springer-Verlag (1997) 327–343.
7. Vassilev, V.K., Miller, J.F., Fogarty, T.C.: Digital circuit evolution and fitness landscapes. In: Proceedings of the Congress on Evolutionary Computation. Volume 2., Piscataway, NJ, IEEE Press (1999) 1299–1306.
8. Wright, S.: The roles of mutation, inbreeding, crossbreeding and selection in evolution. In Jones, D.F., ed.: Proceedings of the 6th International Conference on Genetics. Volume 1. (1932) 356–366.
9. Kauffman, S.A.: Adaptation on rugged fitness landscapes. In Stein, D., ed.: Lectures in the Sciences of Complexity. SFI Studies in the Sciences of Complexity. Addison-Wesley, Reading, MA (1989) 527–618.
10. Manderick, B., de Weger, M., Spiessens, P.: The genetic algorithm and the structure of the fitness landscape. In Belew, R.K., Booker, L.B., eds.: Proceedings of the 4th International Conference on Genetic Algorithms. San Mateo, CA, Morgan Kaufmann (1991) 143–150.
11. Mitchell, M., Forrest, S., Holland, J.: The royal road for genetic algorithms: Fitness landscapes and ga performance. In Varela, J., Bourgine, P., eds.: Proceedings of the 1st European Conference on Artificial Life. Cambridge, MA, MIT Press (1991) 245–254.
12. Palmer, R.: Optimization on rugged landscapes. In Perelson, A., Kauffman, S., eds.: Molecular Evolution on Rugged Landscapes. Volume IX of SFI Studies in the Sciences of Complexity. Addison-Wesley, Reading, MA (1991) 3–25.
13. Wagner, G.P., Altenberg, L.: Complex adaptations and the evolution of evolvability. Evolution 50 (1995) 967–976.

14. Wolpert, D.H., Macready, W.G.: No free lunch theorems for optimization. IEEE Transactions on Evolutionary Computation **1** (1997) 67–82.
15. Stadler, P.F., Seitz, R., Wagner, G.P.: Evolvability of complex characters: Dependent fourier decomposition of fitness landscapes over recombination spaces. Technical Report 99-01-001, Santa Fe Institute (1999).
16. Vassilev, V.K., Fogarty, T.C., Miller, J.F.: Smoothness, ruggedness and neutrality of fitness landscapes: from theory to application. In Ghosh, A., Tsutsui, S., eds.: Theory and Application of Evolutionary Computation: Recent Trends. Springer-Verlag, London (2000) In press.
17. Miller, J.F., Job, D., Vassilev, V.K.: Principles in the evolutionary design of digital circuits. Journal of Genetic Programming and Evolvable Machines **1** (2000).
18. Reidys, C.M., Stadler, P.F.: Neutrality in fitness landscapes. Technical Report 98-10-089, Santa Fe Institute (1998).
19. Stadler, P.F.: Spectral landscape theory. In Crutchfield, J.P., Schuster, P., eds.: Evolutionary Dynamics — Exploring the Interplay of Selection, Neutrality, Accident and Function. Oxford University Press, New York (1999).
20. Kimura, M.: Evolutionary rate at the molecular level. Nature **217** (1968) 624–626.
21. King, J.L., Jukes, T.H.: Non-darwinian evolution. Science **164** (1969) 788–798.
22. Ohta, T.: Slightly deleterious mutant substitutions in evolution. Nature **246** (1973) 96–97.
23. Ohta, T.: The nearly neutral theory of molecular evolution. Annual Review of Ecology and Systematics **23** (1992) 263–286.
24. Huynen, M.A., Stadler, P.F., Fontana, W.: Smoothness within ruggedness: The role of neutrality in adaptation. Proceedings of the National Academy of Science U.S.A. **93** (1996) 397–401.
25. Huynen, M.A.: Exploring phenotype space through neutral evolution. Journal of Molecular Evolution **43** (1996) 165–169.
26. Banzhaf, W.: Genotype-phenotype-mapping and neutral variation — a case study in genetic programming. In Davidor, Y., Schwefel, H.P., Männer, R., eds.: Parallel Problem Solving from Nature III. Berlin, Springer-Verlag (1994) 322–332.
27. Harvey, I., Thompson, A.: Through the labyrinth evolution finds a way: A silicon ridge. In Higuchi, T., Iwata, M., Liu, W., eds.: Proceedings of the 1st International Conference on Evolvable Systems. Berlin, Springer-Verlag (1996) 406–422.
28. Miller, J.F.: An empirical study of the efficiency of learning boolean functions using a cartesian genetic programming approach. In Banzhaf, W., Daida, J., Eiben, A.E., Garzon, M.H., Honavar, V., Jakiela, M., Smith, R.E., eds.: Proceedings of the 1st Genetic and Evolutionary Computation Conference. Volume 2., San Francisco, CA, Morgan Kaufmann (1999) 1135–1142.
29. Miller, J.F., Thomson, P.: Cartesian genetic programming. In: Proceedings of the 3rd European Conference on Genetic Programming. Berlin, Springer-Verlag (2000).
30. Schwefel, H.P.: Numerical Optimization of Computer Models. John Wiley & Sons, Chichester, UK (1981).
31. Bäck, T., Hoffmeister, F., Schwefel, H.P.: A survey of evolutionary strategies. In Belew, R., Booker, L., eds.: Proceedings of the 4th International Conference on Genetic Algorithms. San Francisco, CA, Morgan Kaufmann (1991) 2–9.
32. Mühlenbein, H., Schlierkamp-Voosen, D.: The science of breeding and its application to the breeder genetic algorithm (BGA). Evolutionary Computation **1** (1993) 335–360.
33. Vassilev, V.K., Fogarty, T.C., Miller, J.F.: Information characteristics and the structure of landscapes. Evolutionary Computation **8** (2000) 31–60.

Genetic Algorithm-Based Design Methodology for Pattern Recognition Hardware

Moritoshi Yasunaga[1], Taro Nakamura[2], Ikuo Yoshihara[3], and Jung H. Kim[4]

[1] Institute of Information Sciences and Electronics, University of Tsukuba, Tsukuba, Ibaraki 305-8573, JAPAN
[2] Doctoral Program of Engineering, University of Tsukuba, Tsukuba, Ibaraki 305-8573, JAPAN
[3] Faculty of Engineering, Miyazaki University, Miyazaki 889-2155, JAPAN
[4] Center for Advanced Computer Studies, University of Louisiana, Lafayette, Louisiana, 70504, USA

Abstract. In this paper, we propose a new logic circuit design methodology for pattern recognition chips using the genetic algorithms. In the proposed design methodology, pattern data are transformed into the truth tables and the truth tables are generalized to adapt the unknown pattern data. The genetic algorithm is used to choose the generalization operators. The generalized, or evolved truth tables are then synthesized to logic circuits. Because of this data direct implementation approach, no floating point numerical circuits are required and the intrinsic parallelism in the data is embedded into the circuits. Consequently, high speed recognition systems can be realized with acceptable small circuit size. We have applied this methodology to the face image recognition and the sonar spectrum recognition tasks, and implemented them onto the developed FPGA-based reconfigurable pattern recognition board. The developed system demonstrates high recognition accuracy and much higher processing speed than the conventional approaches.

1   Introduction

We propose the data direct implementation (DDI) approach for practical pattern recognition tasks, in which the recognition circuits are designed and synthesized from the pattern data directly. The DDI enables high speed pattern recognition VLSIs with acceptable small circuit size because of two following reasons.

1. Intrinsic high parallelism in the pattern data set is automatically embedded into the circuits.
2. No complex distance calculation, which is indispensable in the conventional pattern recognition algorithm, is required.

In order to realize the DDI concept, we propose the genetic algorithm-based hardware design methodology based on the pattern data as follows.

1. Sampled pattern data are transformed into the truth tables.
2. The truth tables are evolved to adapt (to recognize) the unknown pattern data. The genetic algorithm [1] is used to choose generalization operators for the evolution.

J. Miller et al. (Eds.): ICES 2000, LNCS 1801, pp. 264–273, 2000.

3. The central circuit of the recognition chip is synthesized from the evolved truth tables and implemented onto the reconfigurable chips.

We name this methodology LoDETT (Logic Design using the Evolved Truth Table). Pioneer works in the evolvable hardware for image recognition have been already reported [2] [3]. This paper propose another promising approach to the real pattern recognition based on the DDI approach with the genetic algorithm-based design methodology LoDETT. We have applied the LoDETT to the face image recognition and sonar spectrum recognition hardware design. The designed (synthesized) circuits have been implemented onto the newly developed FPGA-based reconfigurable pattern recognition board.

In this paper, we describe the newly proposed design methodology LoDETT and its application to the sonar spectrum and the face image recognition hardware.

2   LoDETT

2.1   Fundamental Process Flow and Its Theoretical Background (Application of GA to Parzen Density Estimate)

Fig. 1 shows the process flow of LoDETT. The pattern database stores the pattern data with their categories. Two kinds of truth tables are formed from the stored pattern data (see Fig. 1(a)). Half of the database is used for fundamental truth tables (F-Tables), and the remaining half for training truth tables (T-Tables). For every category, both F-Table and T-Table are made. Each pattern datum is randomly selected and put into either F-Table or T-Table. Rows in the tables of Fig. 1(a) represent the binary-coded pattern data.

Next, the genetic algorithm (GA) is applied to each F-Table in order to put "don't cares" (*'s) in appropriate places using the T-Tables. That is, the F-Table evolves into the E-Table (i.e., the evolved table) by adapting itself to the training pattern data of the T-Tables (see Fig. 1(b)). Through the evolutionary process, some rows in F-Tables are expected to be changed, to match unknown pattern data whose categories are not identified. In other words, through the evolutionary process, each E-Table is expected to obtain generalization capability to categorize the unknown patterns; i.e., the features behind each category's patterns are expected to be embedded in the E-Table automatically. In the recognition process to categorize the unknown pattern data, we find the E-Table that has the largest number of rows matching the unknown pattern datum, and the category of the found E-Table is regarded as that of the unknown pattern datum.

VLSI implementation starts once the E-Tables are made. Hardware implementation of the truth-tables is easily carried out with AND gates through the AND-plane expansion (Fig. 1(c)). Counters and a maximum detector circuit are added to the AND-plane (Fig. 1(d)). The roles of the counters and the maximum detector are to find the E-Table which has the largest number of rows matching the unknown pattern datum (details are described in the next section). Finally, all circuits are implemented onto FPGA chip (Fig. 1(e)).

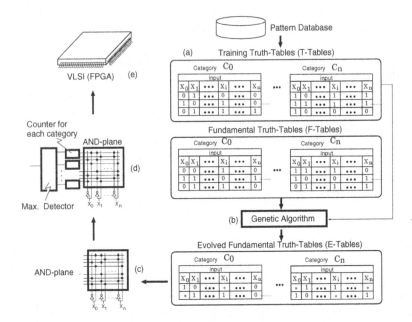

Fig. 1. LoDETT (Logic Design with Evolved Truth-Table).

Theoretical background of LoDETT derives from the Parzen density estimate [4] which is one of non-parametric pattern recognition algorithms and follows the Bayes decision rule. In the Parzen density estimate (Fig. 2), a kernel function, which has a region R, is defined by the diameter D centered at each sampled pattern datum point in the pattern space. Number of kernel functions which the unknown pattern falls into is counted in each category, and the category which has the largest number is chosen as the unknown pattern's category (answer).

The Parzen density estimate gives the best decision if the optimal kernel size R can be chosen at each sampled pattern point to reproduce the class-conditional density of each category. However, no general procedure has been found to decide the optimal kernel size yet. This difficulty impedes the application of the Parzen density estimate to the practical problems. Changing the kernel size R is equivalent to replacing some fields in the truth table to "don't cares" In LoDETT, the optimal kernel size of each sampled pattern point, that is, the optimal positions of "don't cares" in each row in the truth table are searched and chosen by using GA.

2.2　Evolution of Truth Table

The evolution of a row in the F-Table is shown in Fig. 3. Each row in the F-Tables can be transformed into one AND gate (Fig. 3(a)). In this case, the AND gate has six inputs corresponding to the row: three inputs X_2, X_3, X_6 are positive and the remaining ones are negative. This gate can be activated only by the input $\{X_1, X_2, X_3, X_4, X_5, X_6\} = \{0, 1, 1, 0, 0, 1\}$. However, if this row changes

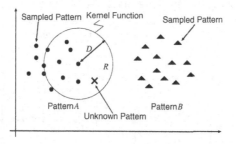

Fig. 2. Parzen Density Estimate (Theoretical Background of LoDETT).

into the row shown in Fig. 3(b) during the evolutionary process, all patterns of $\{X_1, X_2, X_3, X_4, X_5, X_6\} = \{0, 1, *, 0, *, 1\} = (\ \{0, 1, 0, 0, 0, 1\}\ ,\ \{0, 1, 0, 0, 1, 1\}$, $\{0, 1, 1, 0, 0, 1\}$, $\{0, 1, 1, 0, 1, 1\}$) can activate the AND gate. That is, this gate classifies these queries into the same category. How to construct a AND-plane can be well explained using Fig. 3(b). In the figure, in the horizontal line corresponding to the row of $\{X_1, X_2, X_3, X_4, X_5, X_6\} = \{0, 1, *, 0, *, 1\}$, only 4 cross-points of $X_1 = 0, X_2 = 1, X_4 = 0$, and $X_6 = 1$ are connected in the AND-plane. Since the row has $X_3 = *$ and $X_5 = *$, the cross-points of X_3 and X_5 are not connected (i.e., not marked) in the AND-plane. In other words, the input which is *, will not be involved in the AND gate operation, thus will not be connected in the AND-plane.

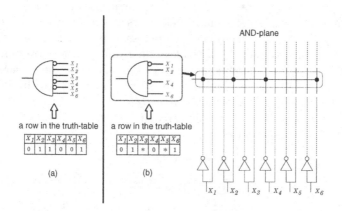

Fig. 3. Evolution of a row in the fundamental truth-table.

Fig. 4 shows the recognition process in the circuits to categorize the unknown pattern. The unknown pattern (query) is put into the AND-plane, i.e., the unknown pattern is broadcast to every AND gate in all categories. If the unknown pattern matches the AND gate, the AND gate is activated (i.e., the output of the AND gate is '1'). This matching test is carried out in parallel in all AND gates.

Each counter counts the number of activated AND gates in the same category, and the category in which the counter outputs the maximum value is detected by the maximum detector and is chosen as the answer, as shown in Fig. 4.

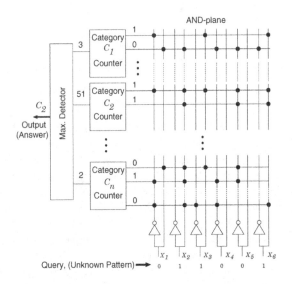

Fig. 4. Recognition Circuits (Recognition Process).

The evolutionary process in LoDETT is carried out in principle as follows. The chromosome consists of a chain of elements which correspond to the rows in the F-Table (Fig. 5). Each locus in the chromosome has either '%' or '&' operator (gene), where the operator '%' keeps the same value in the corresponding field of the table while the operator '&' changes the value into 'don't care'. One of two operators (genes) is randomly selected for each locus in the initial chromosomes. N individuals (chromosomes) are used for each category, and they evolve independently from the other categories' individuals, i.e., there is no interaction among the different categories.

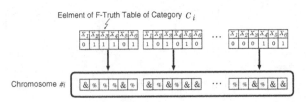

Fig. 5. Chromosomes in GA Process.

The GA operations (selection, crossover, mutation, etc.) are performed on these chromosomes through generations. In each generation, the chromosome is evaluated as shown in Fig. 6. The chromosome is operated onto the F-Table, then we examine how correctly the E-Table can categorize rows (elements) in the T-Tables. Detailed examples of GA operations and fitness are shown in the next section which shows the sonar spectrum and face image recognition hardware designed by LoDETT.

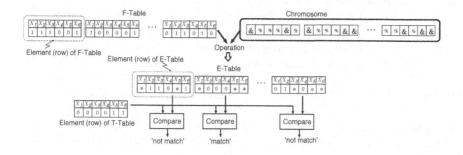

Fig. 6. Evaluation of chromosomes.

3 Face Recognition and Sonar Spectrum Recognition Hardware

3.1 Chromosome Structure

We have applied LoDETT to the face image recognition and the sonar spectrum recognition hardware development. In the design, we used the face image recognition and the sonar spectrum recognition benchmark databases provided by the Olivetti Research Laboratory [1] and the Carnegie Mellon University Computer Science Department [2], respectively. Both benchmark databases are summarized in Table 1.

Table 1. Recognition Problem Benchmark Databases.

Recognition Task	Database Size	Categories	Dimension of Datum	Precision	Source
Face Image	400 images	40 persons	64*	8-bit	Olivetti Res. Lab.
Sonar Spectrum	208 spectra	'Rock' and 'Metal'	60	8-bit	Carnegie Mellon Univ.

* after preprocessing (sampled pattern)

Chromosome structure and genes used in the design are shown in Fig. 7. In the face recognition, each original face image is preprocessed to the sampled image of 64 (8 x 8) pixels. This dimensional reduction (sampling process) is

---

[1] http://www.camorl.co.uk/facedatabase.html

[2] http://www.boltz.cs.cmu.edu/

generally carried out in the conventional image recognition systems to make the systems more robust against image distortion (rotation, zooming, translation, etc.). Each preprocessed face image corresponds to a row in the truth tables. Thus, one row in the truth table is composed of 64 segments x 8 fields (bits). Each segment in the row corresponds to a locus in the chromosome.

In the sonar spectrum recognition, the spectrum data can be directly transformed into the truth tables without any preprocess, and the rows in the truth tables consist of 60 segments x 8 fields (8 bits).

Because of 8-bit precision (8 fields in the locus), there exist 2^8 genes which can be used in the GA operation. However, using all of them causes the standstill of the evolution because of the diversity of genes. Therefore, in the GA operation, a reduced set of genes is used for both tasks. We used the same set of nine genes ('a' - 'i'), as shown in Fig. 7. The gene 'a' keeps the bits (values) in the segment. By the genes 'b' - 'h', bits in the segment are gradually replaced to 'don't cares' from the least significant bit (LSB) to the most significant bit (MSB). The gene 'i' replaces all bits in the segment to 'don't cares' (no AND gate is required to the gene 'i' in the hardware implementation). In the initial generation, each locus is occupied with randomly selected one of nine genes.

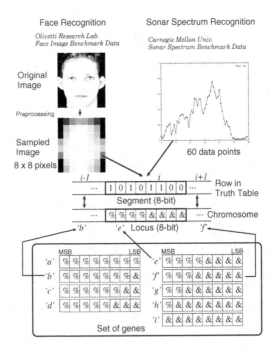

Fig. 7. Chromosome structure and set of genes.

3.2 GA Operation

The simple GA operations (i.e., selection, crossover, mutation) are executed on N chromosomes in each category through generations. In the crossover operation, we apply two-point crossover to the chromosomes. The two-point crossover operation can be simply applied to the sonar spectrum recognition problem because the 60 data points in the spectrum stand in a line. However, in the face image recognition problem, we have to consider the topological relations among pixels in the image. There are four crossover types corresponding to the two crossover points p and q in the image data as shown in Fig. 8, where dashed genes are exchanged between the two chromosomes. In every generation, we randomly choose one of the crossover types as well as pairs of two chromosomes and the crossover points.

In the mutation operation, some loci in all chromosomes are randomly chosen and the genes in them are exchanged to one of the other genes.

Each chromosome is evaluated by using the E-Table (see Fig. 6) and we used the following fitness F in both tasks:

$$F = \frac{G}{1 + H} \qquad (1)$$

$$G = \sum_{i=1}^{n_G} \left(\frac{1}{1 + as_i} \right) \qquad (2)$$

$$H = \sum_{i=1}^{n_H} h, h = \begin{cases} 1 : s_i = 0 \\ 0 : otherwise \end{cases} \qquad (3)$$

where G and H are sub-fitness calculated for the E-Table in the same category and in the different categories, respectively. n_G and n_H are the number of rows in the T-Tables in the same category and the different categories, respectively. s_i is the number of mismatched segments (pixels or data points in the spectrum) in the matching (comparing) test for each row i in the T-Table (see Fig. 6). Constant a is the weighting parameter and it was set to 4.0 in the design.

We used 30 ($=N$) individuals (chromosomes) for each category, and set the crossover rate and the mutation rate to 1.0, and 0.1, respectively. We evolved each population until 1,000 generations with the elitist reservation selection strategy. After the GA operation, the chromosome of the highest fitness is used to design the AND-gate plane in each category. In each category, it took about 16 minutes and 12 minutes in the face and sonar spectrum recognition problems, respectively, for the GA operation to evolve the truth table on the Pentium 166 MHz PC.

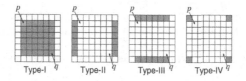

Fig. 8. Four crossover types for image data.

3.3   Performance

Table 2 lists the performance of the developed hardware. We obtained high recognition accuracy in both tasks comparing with the results for the conventional approaches [5] [6]. These results shows that the kernel size R for each datum was chosen very well to reproduce the class-conditional density in the LoDETT.

In spite of the high complexity of the pattern data, the synthesized circuit sizes are small enough to be implemented onto the current FPGA chips widely spread in the market. Reason of this complexity reduction is explained as follows. Through the evolutionary process, common features in the pattern data are extracted and gene 'i' (Fig. 7), which means no AND gate is needed, is operated onto them. Fig. 9 shows an example of a chromosome element which corresponds to an image datum. As shown in the figure, the gene 'i' is operated onto many pixels having common features (back wall and a part of hair, neck and forehead, etc.). This feature extraction in the evolutionary process brings tremendous reduction of circuit complexity.

With the 25 MHz oscillator in the developed hardware, one unknown pattern is recognized within 150 ns including serial/parallel data conversions in the interface circuits in the chip.

Table 2. Performance.

Recognition Task	Recognition Accuracy (%)	Circuit Complexity (2-NAND gates)	Operation Speed (ns)
Face Image	94.7 (Ave.) (90 (Ave.) by eigen face rule [5])	250,604 (Ave.)	< 150
Sonar Spectrum	91.7 (Max.) 83.0 (Ave.) (90.4(Max.) by BackProp. [6])	4,134 (Ave.)	< 150

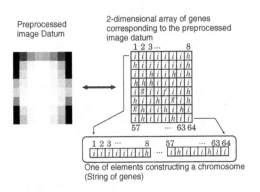

Preprocessed image Datum

2-dimensional array of genes corresponding to the preprocessed image datum

One of elements constructing a chromosome (String of genes)

Fig. 9. Feature extraction through out the evolution process.

3.4   Reconfigurable Pattern Recognition Board

We have developed the FPGA-based reconfigurable pattern recognition board (Fig. 10) onto which the circuits designed by LoDETT are implemented. On the board, seven FPGA chips (Xilinx XCV300-6GB432), data buffer memory, and interface circuits for PC and real time signal input are mounted. Five FPGA chips out of seven, which are equivalent to 1,500k gates in total, can be used for the recognition circuits (Fig. 3), so that the board can be used for many large practical applications.

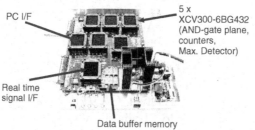

PC I/F

5 x
XCV300-6BG432
(AND-gate plane,
counters,
Max. Detector)

Real time
signal I/F

Data buffer memory

Fig. 10. FPGA-based reconfigurable pattern recognition board.

4   Conclusion

We have proposed the logic design methodology using the GA for the pattern recognition hardware. In the design, truth tables transformed from pattern database are evolved in order to obtain generalization capability, and the evolved truth tables are synthesized into logic gates (AND-plane). We have applied the proposed methodology to the sonar spectrum recognition and the face image recognition hardware design and implemented them onto the FPGA-based reconfigurable pattern recognition board. The developed hardware is composed of less than 100k gates and one unknown pattern can be recognized within 150 ns at high recognition accuracy rate more than 91 %, both in the face image and the sonar spectrum recognition problems.

Acknowledgment

The authors would like to thank the Olivetti Research Laboratory and the Carnegie Mellon University Compute Science Department for their provision of the benchmark databases. This research is supported in part by the 1999 MESSC grant in Japan and the 1999 Japan Science and Technology Corporation grant.

References

1. Goldberg, D.E.: Genetic Algorithms in Search, Optimization, and Machine Learning. Addison-Wesley (1989)
2. Iwata, M., Kajitani, I., Yamada, H., Iba, H., Higuchi, T.: A Pattern Recognition System Using Evolvable Hardware. Lecture Notes in Computer Science, Vol. 1141. Springer-Verlag, (1996) 761–770
3. Torresen, J.: A Divide-and-Conquer Approach to Evolvable Hardware. Lecture Notes in Computer Science, Vol. 1478. Springer-Verlag, (1998) 57–65
4. Fukunaga, K.: Introduction to Statistical Pattern Recognition. Academic Press (1990)
5. Turk, M., Pentland, A.: Eigenfaces for recognition. J. Cognitive Neuroscience. Vol.3 (1991) 71–86
6. Gorman, R.P., Sejnowski, T.J.: Analysis of Hidden Units in a Layered Network Trained to Classify Sonar Targets. J. Neural Networks. Vol.1 (1988) 75–89

A Flexible Model of a CMOS Field Programmable Transistor Array Targeted for Hardware Evolution

Ricardo Salem Zebulum   Adrian Stoica   Didier Keymeulen

Jet Propulsion Laboratory
California Institute of Technology
Pasadena, CA 91109
ricardo@brain.jpl.nasa.gov

Abstract. This article focuses on the properties of a fine grained re-configurable transistor array currently under test at the Jet Propulsion Laboratory (JPL). This Field Programmable Transistor Array (FPTA) is integrated on a Complementary Metal-Oxide Semiconductor (CMOS) chip. The FPTA has advantageous features for hardware evolutionary experiments when compared to programmable circuits with a coarse level of granularity. Although this programmable chip is configured at a transistor level, its architecture is flexible enough to implement standard analog and digital circuits' building blocks with a higher level of complexity. This model and a first set of evolutionary experiments have been recently introduced. Here, the objective is to further illustrate its flexibility and versatility for the implementation of a variety of circuits in comparison with other models of re-configurable circuits. New evolutionary experiments are also presented, serving as a basis for the authors to devise an improved model for the FPTA, to be fabricated in the near future.

1 Introduction

In the context of Evolvable Hardware (EHW), researchers have been using programmable circuits as platforms for their experiments. These programmable circuits are divided into two classes, Field Programmable Gate Arrays (FPGAs) and Field Programmable Analog Arrays (FPAAs). The former is intended to be used in the digital domain, and the latter in the analog domain. Both the *FPGAs* and the *FPAAs* commercially available have a coarse granularity, which may restrict the potential of evolutionary design to the achievement of well-known topologies that are possible with such components. In order to overcome this problem, a fine-grained model of a programmable circuit is being tested at JPL. This programmable device, called the FPTA, provides the benefits of reconfiguration at the transistor level, the use of CMOS technology, and the possibility of synthesizing circuits in the analog

J. Miller et al. (Eds.): ICES 2000, LNCS 1801, pp. 274–283, 2000.

and in the digital domain. A first version of this programmable chip has already been successfully employed in the evolution of a computational circuit [11].

The purpose of this paper is twofold: to highlight the advantages of a fine-grained reconfigurable architecture that is also able to provide circuits with higher levels of granularity as building blocks for evolution; and to present preliminary evolutionary experiments involving the synthesis of circuits whose main characteristics are analyzed in different domains. The capacity of mapping circuits of higher complexity, while still being programmable at the transistor level, gives the designer the possibility to choose the most adequate building blocks to be manipulated by the evolutionary algorithm. Additionally, the evolution of circuits in different analysis' domains shows the flexibility of this model to accomplish the synthesis of a wide variety of electronic circuits.

This paper is divided in the following way: section 2 briefly summarizes some models of programmable circuits; section 3 reviews the basic features of the programmable device developed at JPL; section 4 describes the mapping of some standard circuit building blocks into this FPTA; section 5 presents two evolutionary experiments, where the FPTA is used as a model to synthesize circuits in different domains of analysis. An enhanced model of the FPTA is also introduced in this section. Section 6 concludes the work, and describes future developments and applications of this reconfigurable platform.

2   Reconfigurable Circuits

Field Programmable Analog Arrays and *Field Programmable Gate Arrays* promise to establish a new trend in electronic design, where a single device now has the flexibility to implement a wide range of electronic circuits. While FPGAs have been developed for applications in the domain of digital signal processing and re-configurable computing, most FPAA models are being developed for applications in programmable mixed-signal circuits and filtering. In addition to the intrinsic flexibility of these devices, which confers advantageous features to standard electronic design, FPGAs and FPAAs are also the focus of research in the area of *self programmable* systems [3][7][12]. Particularly, genetic algorithms are the main agents employed to promote the property of automatic configuration.

Nevertheless, there are many issues that should be addressed prior to using a programmable circuit in the context of artificial evolution. Perhaps, the most important of these issues is that of granularity of the programmable chip. The devices presented so far can be divided into two classes, coarse grained and fine grained. While the former uses more complex circuits, such as digital gates and operational amplifiers, as the building blocks for the design process, the latter is configurable at a lower level, usually at the transistor level.

Most FPGA models consist of an arrangement of cells that perform digital logic, such as basic gates, multiplexers and flip-flops [7]. The user can configure the cells' connections and, in some models, their functionality. Many surveys of FPGA models can be found in the literature [9]; therefore, this section focuses on the description of FPAAs, whose development has occurred more recently. Five models are described here.

2.1   TRAC – Totally Reconfigurable Analog Hardware

This device is a coarse grained FPAA, consisting of twenty operational amplifiers laid out as two parallel sets of ten inter-connectable amplifiers [2]. The Zetex programmable chip has been used in evolutionary experiments that targeted the synthesis of computational functions [2], rectifiers, and others [8]. The main limitation of the use of this chip in EHW is that evolution is constrained to the arrangement of high level conventional building blocks to compose the final topology, not being able to arrive at any novel design.

2.2   Motorola Field Programmable Analog Array (MPAA020)

The analog resources of the coarse grained MPAA020 [4] are distributed along 20 nearly identical cells. Each cell includes the following hardware resources: one CMOS operational amplifier; one comparator; capacitors; programmable switches; and SRAM. This is the hardware required for the design of switched capacitor based circuits. This FPAA is configured by a bitstring of 6864 bits that control the circuit connectivity, capacitors' values and other features. This programmable chip has been used as a platform for the evolution of filters, oscillators, amplifiers, and rectifiers [13]. The MPAA020 provides more resources for evolution, since programmable capacitors and resistors (implemented through the switched capacitor effect) are now included on chip. However, since the basic building blocks are fixed to Operational Amplifiers (OpAmp) topologies, there is little room for evolution to arrive at novel designs. Another limitation of the current version of this chip is that it is not a transparent architecture, preventing simulation/analysis of evolved circuits.

2.3   Palmo (University of Edinburgh)

The Palmo development system [1] is BiCMOS technology coarse-grained FPAA chip that works on pulse-based techniques. The chip architecture consists of an array of programmable cells that perform the functionality of integrators. One of the design goals is to use the chip in EHW experiments, employing GAs to determine the integrators' interconnectivities and their gains.

2.4   Evolvable Motherboard (University of Sussex)

The Evolvable Motherboard (EM) [5] is a research tool designed for access at low level granularity. This programmable board allows different components to be plugged in as basic circuit elements. The board is organized as a matrix of components, where digitally controlled analog switches allow row/column interconnection. In total, approximately 1500 switches are used, giving a search space of 10^{420} possible circuits. The EM has already been used in evolutionary experiments whose objectives were the design of inverter gates, amplifiers and oscillators. In these experiments, bipolar transistors were the components manipulated by the evolutionary system.

2.5 – Lattice Programmable Analog Circuits (PACs)

The Lattice PAC is a coarse-grained programmable analog circuit that was commercially introduced recently [6], intended for applications in filtering, summing/differencing, gain/attenuation, and conversion.   The PAC cell includes instrumentation amplifiers, an active resistance and a programmable capacitor. The major limitation for the application of this chip to EHW is the fact that it is programmed through non-volatile memory cells with limited number of program/erase cycles , around 100,000, insufficient for evolutionary experiments.

The devices previously described have limitations from the point of view of their suitability in implementing *Evolution Oriented Reconfigurable Architectures* (EORA). In the case of the commercial devices (TRAC, MPAA020 and Lattice), the main problems are their coarse granularity and, in some cases, the non-transparent architecture. Coarse granularity constrains evolution to sample designs based on human conceived building blocks. The academic research tools, EM and Palmo, are a board level solution and a coarse grained design respectively. EM has inherent flexibility, since components are added by human plug-in, but will require a decision of what resources would finally go on a VLSI chip.

3   Field Programmable Transistor Array (FPTA)

This section reviews the architecture of the FPTA (more details can be found in [10]) This transistor array combines the features of fine granularity, being programmable at the transistor level, with the architectural versatility for the implementation of standard building blocks for analog and digital designs.

The Field Programmable Transistor Array is a fine-grained reconfigurable architecture targeted for Evolvable Hardware experiments. The basic components are MOS transistors, integrated on a chip using 0.5-micron CMOS technology [10]. As both analog and digital CMOS circuits ultimately rely on functions implemented with transistors, the FPTA appears as a versatile platform for the synthesis of both analog and digital (and mixed-signal) circuits. Moreover, as will be demonstrated in the next section, the FPTA architecture is also flexible enough to provide more complex building blocks to the evolutionary system.

The FPTA cell is an array of transistors interconnected by programmable switches. The status of the switches (ON or OFF) determines a circuit topology and, consequently, a specific response. Figure 1 illustrates an example of an FPTA cell consisting of 8 transistors and 24 programmable switches. This cell consists of PMOS and NMOS transistors, and switch based connections. As will be further described in this paper, this cell architecture allows the synthesis of both standard and novel topologies for some basic circuits. According to the outcome of the experiments being currently performed, a slightly different architecture may be proposed for the next version of the FPTA.

The programmable switches are implemented with transistors acting as transmission gates. The switches can either be programmed for the implementation of minimal and maximal resistance values, by applying 0V and 5V at the transistors' gates, or they can be programmed to implement intermediate resistance values, by applying values between 0V and 5V at the transistors' gates. Preliminary experiments suggest that the use of intermediate values for switching control generates a smoother fitness landscape [11].

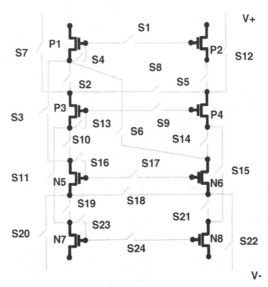

Fig. 1 – Schematic of a FPTA cell consisting of 8 transistors and 24 switches.

4   Mapping Standard Circuit Building Blocks Onto the FPTA

This section describes recent experiments performed with the FPTA chip, referring to the mapping of some standard circuit building blocks onto the FPTA. The objective is to show that the architecture depicted in Figure 1 is flexible enough to implement basic structures used in analog and digital design. We depict here the implementation of three cells: a basic common-source amplifier; a transconductance amplifier; and an AND digital gate. These are examples of circuits that can be optionally used as building blocks for the evolution of analog and digital circuits.

Figure 2.A depicts the FPTA configuration for a common source amplifier implementation. This circuit is employed to provide voltage gain. This structure can be easily mapped into one cell of the transistor array. As shown in the figure below, transistors P3 and P4 form a current mirror pair that works as the active load for the amplifying transistor, N6. The current level is determined by the input voltage *Vbias*, applied at the gate of the transistor N5, which serves as an active resistance for the current mirror. Figure 2.B compares the DC transfer function of these circuits in two cases: simulation and real implementation. In the former, we use PSPICE to simulate the amplifier without switches implementing the connections; in the latter, we get real data from the amplifier implementation.

From Figure 2 we can observe that the transfer curves of the two versions of the circuit have the shape of an inverter amplifier. They only differ in their higher saturation values, 5V for the simulated circuit and 4.6V for the real one. This difference is due to the fact that the circuit is very sensitive to the DC operating point, given by the bias current.

Another circuit building block that can be easily mapped into the FPTA is the transconductance amplifier, whose schematic is displayed in Figure 3. Also shown in Figure 3 is the graph comparing the DC transfer functions between the tranconductance amplifier implemented in the FPTA and a simulated version of the same with no switches.

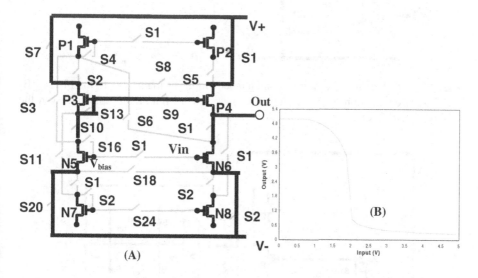

Fig. 2 – (A) - Schematic of the mapping of a basic common source amplifier onto the FPTA. (B) - DC transfer function of the common source amplifier in simulation (full line) and one implemented in the FPTA (dotted line).

Figure 4 displays the schematic of a digital gate mapped into two cells of the FPTA. This digital gate performs NAND/AND operations over the inputs *In1* and *In2*, as shown in the figure below. This circuit was implemented in the FPTA, and worked as expected.

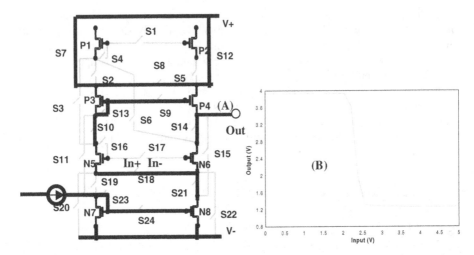

Fig. 3 – (A) - Schematic of a transconductance amplifier mapped onto the FPTA; (B) - DC transfer function of the circuit in simulation (full line) and implemented in the FPTA (dotted line).

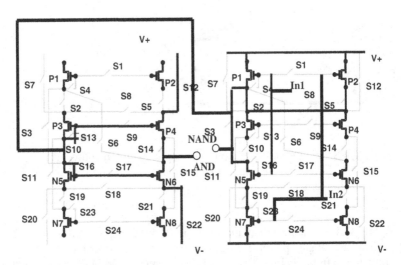

Fig. 4 – Schematic of an AND gate mapped into two FPTA cells.

5   Evolutionary Experiments

This section describes new evolutionary experiments carried out using the same FPTA model in the *PSPICE* simulator: the evolution of an amplifier and of a band-pass filter respectively. We employed a standard Genetic Algorithm with binary representation, 70% one-point crossover ratio and 4% mutation ratio.

In the first experiment, the objective was to synthesize a circuit with a DC transfer characteristic typical of an amplifier, similar to the ones illustrated in the last section.

The chromosomes were represented by 24 bit strings, where each bit maps the state (opened or closed) of one switch. The following fitness evaluation function was used:

$$\text{Fitness} = \text{Max}_{i=1}^{n-1} |V(i+1) - V(i)| \qquad (1)$$

Where $V(i)$ describes the circuit output voltage as a function of the DC analysis index i, which spans the swept range of the input signal, i.e., from 0 to 5 Volts. This evaluation function aims to identify the maximum voltage gradients between consecutive input voltage steps, the larger the gradient, the larger the amplifier gain will be [12]. This fitness evaluation function does not impose a DC operating point for the amplifier. Figure 5 depicts the schematic of the evolved circuit, together with the DC transfer responses achieved in the FPTA implementation and in simulation.

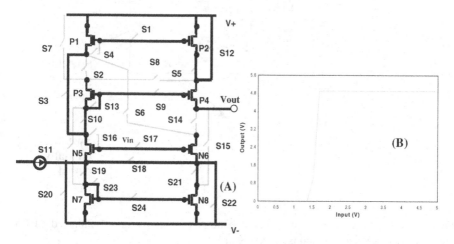

Fig. 5 – (A) - Schematic of an evolved amplifier using the FPTA model; (B) - comparison of the DC characteristic displayed by the simulated (full line) and implemented (dotted line) versions of the amplifier.

This result compares very well with the one reported in [12], where the same fitness function was used, but the basic elements for evolution were bipolar transistors and resistors. While around 10^4 evaluations were needed to obtain a similar DC transfer function in [12], only 900 evaluations (30 individuals along 30 generations in only one GA execution) were necessary using the FPTA. In addition, the evolved circuit shown above can be directly implemented in a CMOS reconfigurable chip.

The second evolutionary experiment targeted the synthesis of a band-pass filter for an AM band, with a range that goes from 150kHz to 1MHz. This experiment served as a testbed for an improved cell model for the FPTA. Since the basic FPTA cell does not include large capacitances (only low-valued parasitic capacitances), we included external capacitances in the original cell, based on a previously evolved filter [14]. This new model also included additional switches, which propagated the input signal to other points in the FPTA. Instead of using values of 0V and 5V to control the switches, intermediate values of 1.5V and 3.5V were employed in this experiment. It has been verified that all of the above mentioned new features improved the performance of the FPTA for filter design. Figure 6 depicts the evolved filter and its frequency response.

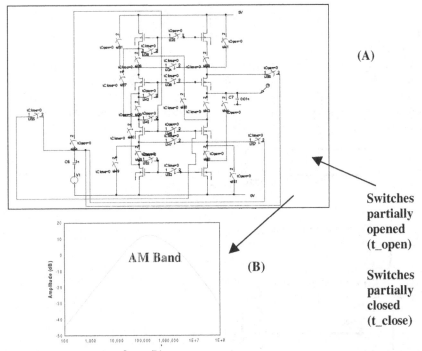

Fig. 6 – (A) - Schematic of an evolved band-pass filter using an enhanced FPTA model; (B) – Frequency response (Axis X in Hz).

The new features can be observed from the schematic above: output and decoupling capacitances were added; and multiple paths from the input signal to the main circuit were allowed. It may be noted that the Genetic Algorithm did not start from scratch in this application because both capacitors' positions and values were known a priori from another evolved design [14]. However, the final frequency response is well centered within the AM band, providing a gain of around 15 dB (typical of human made design for this circuit). In this experiment, a successful GA execution required the sampling of around 10^4 individuals.

6   Conclusions

This paper reviewed the basic features of the programmable chip under test at JPL, the FPTA. We described reconfigurable chips used in EHW. Contrasting with most of these devices, the FPTA provides the benefits of programmability at the transistor level. New experiments were described in this paper, showing that the FPTA architecture, though being fine grained, is flexible enough to map basic analog and digital circuits building blocks. We reported new evolutionary experiments, which have given us support for developing an enhanced design of the FPTA, which will include new switches and programmable capacitors.

Acknowledgements

This research was performed at the Center for Integrated Space Microsystems, JPL, California Inst. of Technology and was sponsored by the Defense Advanced Research Projects Agency (DARPA) under the Adaptive Computing Systems Program managed by Dr. Jose Munoz.

References

1. Hamilton, A., Papathanasiou, K., Tamplin, M. R., Brandtner, T.,, "PALMO: Field Programmable Analog and Mixed-Signal VLSI for Evolvable Hardware", in Proceedings of the Sec. Int. Conf. on Evolvable Systems. M.Sipper, D.Mange and A. Pérez-Uribe (editors), vol. 1478, pp. 335-344, LNCS, Springer-Verlag, 1998.
2. Flockton and Sheehan, " Intrinsic Circuit Evolution Using Programmable Analog Arrays", in Proc. of the Sec. International Conference on Evolvable Systems, M.Sipper, D.Mange and A. Pérez-Uribe (editors), vol. 1478, pp. 144-155, LNCS, Springer-Verlag, 1998.
3. Lohn, J. D., Colombano, S. P., "Automated Analog Circuit Synthesis Using a Linear Representation", in Proc. of the Sec. Int. Conf. on Evolvable Systems:. M.Sipper, D.Mange and A. Pérez-Uribe (ed), vol. 1478, pp. 125-133, LNCS, Springer-Verlag, 1998.
4. Motorola Semiconductor Technical Data, "Advance Information Field Programmable Analog Array 20-Cell Version MPAA020", Motorola, Inc., 1997.
5. Layzell, P., "A New Research Tool for Intrinsic Hardware Evolution", in Proceedings of the Second International Conference on Evolvable Systems. M.Sipper, D.Mange and A. Pérez-Uribe (editors), vol. 1478, pp. 47-56, LNCS, Springer-Verlag, 1998.
6. Lattice Semiconductor Corporation, "ispPAC Handbook, Programmable Analog Circuits", Sep., 1999.
7. Miller, J., Thomson, , P., " Aspects of Digital Evolution: Geometry and Learning", in Proceedings of the Sec. Int. Conf. on Evolvable Systems. M.Sipper, D.Mange and A. Pérez-Uribe (editors), vol. 1478, pp. 25-35, LNCS, Springer-Verlag, 1998.
8. Ozsvald, I., "Short-Circuiting the Design Process: Evolutionary Algorithms for Circuit Design Using Reconfigurable Analog Hardware", Msc Thesis, School of Cognitive and Computer Sciences (COGS), University of Sussex, 1998.
9. Sanchez, E., "Filed Programmable Gate Array (FPGA) Circuits", in *Towards Evolvable Hardware: The evolutionary engineering approach*, pp. 1-18, E. Sanchez and M. Tomassini (editors), Springer-Verlag LNCS 1062, 1996.
10. Stoica, A., "Towards evolvable hardware chips: experiments with a configurable transistor array", Proceedings of the 7th Int. Conf. On Microelectronics for Neural, Fuzzy and Bio-Inspired Systems, Granada, Spain, April, 7-9, IEEE comp. Sci. Press, 1999.
11. Stoica, A., Keymeulen, D., Tawel, R., Salazar-Lazaro, C., Li, W., "Evolutionary Experiments with a fine-grained reconfigurable architecture for analog and digital CMOS circuits" , Proc. of the First NASA DoD Workshop on Evolvable Hardware, pp.76-84, IEEE Computer press., July, 1999.
12. Thompson, A., Layzell, P., Zebulum, R., "Explorations in Design space: Unconventional Electronics Design Through Artificial Evolution", IEEE Trans. on Evolutionary computation, vol. 3, n. 3, pp. 167-196, 1999.
13. Zebulum, R.S., Pacheco, M.A., Vellasco, M., "Analog Circuits Evolution in Extrinsic and Intrinsic Modes", in Proc. of the Sec. Int. Conf. on Evolvable Systems M.Sipper, D.Mange and A. Pérez-Uribe (ed), vol. 1478, pp. 154-165, LNCS, Springer-Verlag, 1998.
14. Zebulum, R.S., Pacheco, M.A., Vellasco, M., "Artificial Evolution of Active Filters: A Case Study", Proc. of the First NASA DoD Workshop on Evolvable Hardware, pp.66-75, IEEE Computer press., July, 1999.

Author Index